WOMEN IN THE EUROPEAN COUNTRYSIDE

Perspectives on Rural Policy and Planning

Series Editors:
Andrew Gilg
University of Exeter, UK
Keith Hoggart
King's College London, UK
Henry Buller
Cheltenham College of Higher Education, UK
Owen Furuseth
University of North Carolina, USA
Mark Lapping
University of South Maine, USA

Other titles in the series

Geographies of Rural Cultures and Societies
Edited by Lewis Holloway and Moya Kneafsey
ISBN 0 7546 3571 6

Mapping the Rural Problem in the Baltic Countryside
Transition Processes in the Rural Areas of Estonia, Latvia and Lithuania
Edited by Ilkka Alanen
ISBN 0 7546 3434 5

Big Places, Big Plans
Edited by Mark B. Lapping and Owen J. Furuseth
ISBN 0 7546 3586 4

Young People in Rural Areas of Europe
Edited by Birgit Jentsch and Mark Shucksmith
ISBN 0 7546 3478 7

Power and Gender in European Rural Development
Edited by Henri Goverde, Henk de Haan and Mireia Baylina
ISBN 0 7546 4020 5

Women in the European Countryside

Edited by

HENRY BULLER
University of Exeter, UK

KEITH HOGGART
King's College London, UK

Routledge
Taylor & Francis Group

LONDON AND NEW YORK

First published 2004 by Ashgate Publishing

Reissued 2018 by Routledge
2 Park Square, Milton Park, Abingdon, Oxon OX14 4RN
605 Third Avenue, New York, NY 10017

First issued in paperback 2021

Routledge is an imprint of the Taylor & Francis Group, an informa business

© Henry Buller and Keith Hoggart 2004

Henry Buller and Keith Hoggart have asserted their right under the Copyright, Designs and Patents Act, 1988, to be identified as the editors of this work.

A Library of Congress record exists under LC control number: 2004012171

Notice:
Product or corporate names may be trademarks or registered trademarks, and are used only for identification and explanation without intent to infringe.

Publisher's Note
The publisher has gone to great lengths to ensure the quality of this reprint but points out that some imperfections in the original copies may be apparent.

Disclaimer
The publisher has made every effort to trace copyright holders and welcomes correspondence from those they have been unable to contact.

ISBN-13: 978-0-815-39910-0 (hbk)
ISBN-13: 978-1-351-14288-5 (ebk)
ISBN 13: 978-1-138-35863-8 (pbk)

DOI: 10.4324/9781351142885

Contents

List of Figures

List of Tables

List of Contributors

Bettina Bock
Assistant Professor of Rural Sociology and Rural Gender Studies, Wageningen University

Henry Buller
Professor of Rural Geography, University of Exeter

Marion Demossier
Senior Lecturer in French Studies, Department of European Studies and Modern Languages, Bath University

Marit S. Haugen
Senior Researcher, Centre for Rural Research, Norwegian University of Science and Technology, Trondheim

Bettina van Hoven
Researcher and Lecturer in Cultural Geography, University of Groningen

Keith Hoggart
Professor of Geography, King's College London

Heide Inhetveen
Professor of Rural Sociology and Rural Gender Studies, University of Goettingen

Ingunn Limstrand
Counsellor, Northern Feminist University, Steigen, Nordland

Mathilde Schmitt
Senior Researcher, Institute of Rural Development, University of Goettingen

Christina Scholten
Centre for Gender Studies, Lund University and Managing Director of Gender Equality, County Administration Board of Kronoberg

Marit Stemland
Counsellor, Northern Feminist University, Steigen, Nordland

Chapter 1

Structures, Cultures, Personalities, Places, Policies: Frameworks for Uneven Development

Keith Hoggart

Introduction

How rapidly is rural society changing? From a gender equality perspective there are good reasons for holding that it is not fast enough. But it is not simply a matter of pace of change, for the nature of change is also critical. On the surface dramatic change might appear to be taking place but how far this penetrates into core values and relationships requires incisive probing and contemplation from a variety of different perspectives. It is in this context that the collection of research essays in this book seeks to contribute to the growing literature on women in rural Europe. When this project was begun, it started with a vision of the existing rural literature that recognized significant advances in the body of evidence and theorization that had developed on rural gender relations, but felt discomforted by a relative absence of broader considerations of the lives of rural women. Although there have been some significant case studies and broader trend additions to the literature since then (e.g. Little, 2002), there is still a need for deepening our understanding of the range of contexts in which women experience life in rural areas. The same point could be made about men, even if the focus of 'gaps' might not be the same. Thus, while the literature has gone a long way to addressing the former invisibility of women in rural studies, there has been little progress made over the shroud that blights our insight on different aspects of men's lives. In much the same way as women were excluded by an emphasis on 'households', 'the workforce' or 'the population' of rural areas, so too were many aspects of men's lives. What research on rural gender relations has brought into focus is similarities and differences in particular spheres of human life, but much remains obscure. For men, for example, there is little knowledge about the actions of rural people as parents, as members of local social networks (save perhaps for what traditionalists might portray as the 'male spheres' of power relations), as individuals seeking personal satisfaction through employment, as adolescents or as senior citizens. That the same points could be made about women provided a key starting point for this collection of chapters. As

Bock and Haan (2004, p.119) have recently noted, little attention has been devoted so far to gender-specific perceptions of the quality of life in rural areas, even if a positive research shift has occurred toward understanding the different ways women and men experience rural life (Baylina and Bock, 2004, p.96). None of this should detract from the massive progress that has been made, with noteworthy contributions extending the range of subjects that have been investigated (e.g. Little, 1997), just as the focus of attention has moved beyond farm-centred economies to wider economic contexts (e.g. Bennett, 2004), and attention to specific population strata has increased (e.g. Jentsch and Shucksmith, 2004).

Accompanying such changes, there has been a strengthening and deepening of conceptualizations and theorizations, as available understandings have been penetrated by new perspectives, which have offered distinctive frameworks from which to interpret (rural) women's lives. In Berg's (2004) words, work in the 'rural women's subordination category', that was a significant component of the research agenda in the 1970s and 1980s, tended to portray women's lives as miserable, and in doing so failed to draw out positive dimensions of both womanhood and women's lives in rural areas. As women's and gender perspectives have taken a more dominant position within research frameworks, a richer appreciation of difference has been forthcoming, although it is arguable that, with the exception of some research on identity, there is still a tendency for commentaries to focus on women's 'subordination'. Given the biases in society that such research is articulating, some aspects of this are inevitable, although there would seem to be a place for more celebratory exploration of women in rural societies.

Certainly it is possible to document improvements in the standing of rural women fairly readily, just as it is easy to offer evidence of biases in society. Yet these are not universal givens, with uneven incidences of advantage and disadvantage for those in rural areas compared with women in other environments, as well as for women compared with men. Even so, recent reports like that of the Commission of the European Communities Directorate-General for Agriculture (2000) paint a clear picture of a continuing pattern of poorer services and weaker paid-work prospects for women compared with men. This does not mean the situation is not changing. For example, Chapman and colleagues (1998, p.25) report that rises in women's wage rates are outpacing those of men, with women's wages in rural Britain now higher than in non-rural areas. Similarly, Demossier (2004, p.46) reports that the percentage of French rural women of working age with paid employment reached the national average in 1997, with 81.9 per cent of rural women having paid-work by 2002, which is only a sliver below the figure for Paris and its suburbs (82.2 per cent), and stands a couple of points above the national figure (79.8 per cent). However, as research on rural women has matured, we have come to appreciate more that such figures only tell part of the story. In the same way that legal provisions only tell us part of the story about equality before the law, so to do statistics only demarcate certain elements of similarity (or dissimilarity) in status. It is not that they 'lie', but that to appreciate their 'realities' we have to read what figures and laws mean 'behind the scenes'. Hence, as Haugen (1994, p.99) pointed out: '... women's formal (judicial) status is today equal with that of men but … the most important negotiations and decisions have to be made in households and communities, and not in the courts'. By blending in-depth qualitative analyses with broader brush survey and statistical studies, investigators have been able both to acknowledge changes in women's position where these seem apparent and probe behind this appearance to deepen understanding. One example of this is identifying negative consequences from changes that seem to offer hope for greater gender equality. Bennett (2004) provides a good illustration, in exploring how a relative economic repositioning of women within household income streams has not led to substantial change in gender relations, with both women and men struggling to cope with the repercussions of economic collapse in coal mining and agriculture. Although couched in a different context, such findings can be placed alongside evidence drawn over a considerable period that women's engagement with paid-work has not necessarily changed intra-household relations in significant ways (e.g. Hillebrand and Blom, 1993). Indeed, with a relatively high dependence on self-employment in rural areas, and with relatively poor support systems for women's employment, it is not surprising to find the conclusion that there has been a tendency amongst women to 'self-exploit' (Inhetveen and Schmitt, 2004); and all this at a time when the media reveal a reluctance to acknowledge women's changing role in rural societies (e.g. Morris and Evans, 2001).

Significantly, of course, whatever changes have occurred, have not become manifest at the same pace, nor in some cases even in the similar direction, with even roughly comparable adjustments having quite different consequences amongst women and across local societies. To make sense of cross-cutting trends, contradictory tendencies and unpredictable outcomes, the detailed investigations that are presented in this book can be located appropriately within the framework provided by Bettina Bock in *Chapter Two*. In exploring the differing positions of women's employment across Europe, this chapter emphasizes how the significant variety that occurs can be understood in terms of the *structural conditions* that pertain in a locality (as seen in the circumstances of local labour markets), in local *cultural conditions* (especially as these related to values pertaining to gender roles), in the circumstances of the *individual* (such as their personal resources), and in the *locations* in which women engage with or experience the interaction of these forces. Added to which, there are the policy frameworks within which women's lives are played out, since these are prone to more rapid short-term change than other key frameworks that affect women's lives; they interact with these other circumstances, giving them new meaning, while restraining or liberating potentialities. As the chapters that follow show, each of these five frameworks, while not operating in isolation from one another, all have something to tell us about the nature of rural women's lives in Europe today.

Structural Contexts

Following on from earlier comments made on the potentially deleterious effects of 1992 Single European Market reforms on creating unfavourable terms for women's integration into labour markets in southern Europe (Hadjimichalis, 1994), Costis

Hadjimichalis (2003) has more recently pointed out that the current emphasis within the European Spatial Development Perspective (ESDP) offers a stress on urban-centred economic organization that gives little acknowledgment to the dissimilar paths that growth in parts of the European South has taken in the recent past. Built within the orientation such European policies provide for rural labour markets are structural restraints that have negative consequences for women:

> Structural changes in the countryside were led by flexible management ventures in which peasant families were engaged in industrial activities as part-time wage earners. ... the combination of autonomy and control, the availability of social networks, the combination of local tradition with innovativeness and flexibility and not least, a gender division of labour that forced women to work with [the] lowest salaries or as 'unpaid family members', constructed a 'new mode of social reproduction' in SE [southern European] regions. (Hadjimichalis, 2003, p.106)

Although it is difficult to generalize across southern Europe as a whole, as contrasts between Portugal and Spain readily attest (Brassloff, 1993), it seems clear that women in the South have particularly restricted employment prospects in rural labour markets. In the Spanish case, for example, while it appears that women in rural areas are more likely to be part of the labour force than the national average, this potentially positive picture rapidly thins when it is noted that 13.4 per cent of rural women work as 'family assistants' (Commission of the European Communities Directorate-General for Agriculture, 2000, p.6). In similar vein, women farmers are disproportionately represented on smaller holdings, as in Portugal (Commission of the European Communities Directorate-General for Agriculture, 2002, p.1), while the negative consequences of seeking work in rural labour markets commonly fall on women. Thus, in Baden-Württemberg, 40 per cent of rural women are reported to hold jobs that require a level of formal qualifications below those that they hold, which compares with a value of 29 per cent for women in urban areas (Commission of the European Communities Directorate-General for Agriculture, 2000, p.6). Indeed, in the very year when the economic activity rate for French rural women matched the national average (in 1997), amongst those aged 15-24 years, 34 per cent of rural women, as compared with 19 per cent of men, were unemployed (Commission of the European Communities Directorate-General for Agriculture, 2000, p.7). Hardly surprisingly in these circumstances, women are reported to be leaving rural areas in large numbers, with the heaviest losses from more peripheral areas, where problems of transport add to the difficulties of securing work (e.g. Lindsay *et al.*, 2003). Thus, in Finland, it is reported that in the most remote rural areas the female proportion of the population aged 25-44 years was only 40 per cent as early as 1988 (Commission of the European Communities Directorate-General for Agriculture, 2000, p.5). In such circumstances the state becomes critical as an employer (e.g. Navarro, 1999; Limstrand and Stemland, 2004); if, that is, the state is intent on pursuing policies that favour of a rural-urban equalization in service standards, which over the last 10-15 years is a policy that has come under considerable strain, even in countries that previously favoured this outcome (e.g. Persson and Westholm, 1993).

Viewed in the context of rural exodus that has left an ageing and declining number of agricultural males in villages, Marion Demossier provides us with an analysis of what this means for rural France in *Chapter Three*. Significantly, this involves a continuing trend of women acting as agents of modernization in the rural world (for an earlier commentary on this effect, see Morin, 1970). Although the loss of many women from rural villages has the potential to intensify social strains, as well as threatening future population stability through impacts on household formation, the introduction of in-migrant families and the necessity of male farmers to seek partners amongst the non-farm population has provided women with an important mediating position between rural and urban France. Here, especially *neo-ruraux* women, in a manner that contradicts messages in the UK literature about in-migrant women seeking more traditional, conservative lifestyles on moving to the countryside (e.g. Hughes, 1997), we find women are very influential in implementing change that is necessary for rural modernization.

Instrumental within this process is a significant change in the occupations of household members, even on farm households, where women have become much more active agents in the paid workforce; bringing important new income sources into farm (and non-farm) households. Marit Haugen draws attention to significant elements of this process in *Chapter Four*, using the Norwegian case to illustrate trends. Significantly here, women who live on farms are found to enjoy farm work, and many would like to do more, but they also value off-farms jobs. Quite apart from social reasons for this latter view, there are obvious economic reasons for wanting off-farm paid-work, since the social security benefits that are available to women who work on-farm only, which extend from pensions, to maternity leave, to unemployment benefit, to sick leave and disability allowances, are not as good as those associated with off-farm jobs. This structural deficit has been associated with farm work for a long time, across a wide range of countries. What intensifies its impact is the difficulties women can face in seeking to secure non-farm jobs, for one of the real problems with rural labour markets is an absolute deficit in job availability (e.g. Breeze *et al.*, 2000), which means that rural women who do not have off-farm work seem more constrained in their ability to obtain even low-paid jobs than men or their non-rural counterparts (Chapman *et al.*, 1998).

Rural Cultures

One consequence this has is to increase the weight that is placed on self-employment as a means of securing work in rural areas. As various chapters in this book show, across a variety of rural settings, the growing involvement of women in paid-work is enhancing the sustainability of rural populations (e.g. Demossier, 2004; Haugen, 2004). However, in the context of tight local labour markets, there are a variety of constraints on women's engagement with paid-work. One substantial hurdle is a value disposition that favours men's involvement in local labour markets:

> The majority of women accepted that given the better pay and greater security accorded to male employment in rural areas, women's aspirations, by necessity, had

to take second place to their male partners or male peer group. These impediments to economic integration were closely bound up with transport and child-care services, both of which were deficient. (Philip and Shucksmith, 2003, p.470)

Little and Panelli (2003) are no doubt correct in asserting that imageries which paint countryside living as 'idyllic' are instrumental in shaping and sustaining patriarchal gender relations. Yet the forces that lie behind patriarchal relationship are also potent in surroundings that most would portray as somewhat less than idyllic. One way to illustrate this point is to explore situations which seem more favourable to women achieving personal autonomy. As various commentators have pointed out, in the sphere of paid-work, there should be more scope for rural self-employment nowadays, given advancements in telecommunications (Clark, 2000; Limstrand and Stemland, 2004). Indeed, it might be anticipated that self-employment should be especially attractive to women in areas where 'rural cultures' favour the retention of traditional gender divisions of household labour (Bock, 2004). If local cultures embody a norm that women have prime responsibility for child-rearing, then poor childcare provision in rural areas is clearly detrimental to women's paid employment (Halliday, 1997),[1] which means that the added flexibility afforded by self-employment offers a potentially easier route into income generation.

But the reality of self-employment in rural areas is that women do not tend to be primary agents in enterprises. Thus, amongst rural entrepreneurs outside agriculture, the employed labour force in Finland in 1994 was 5 per cent for women and 14 per cent for men, with the rate of increase in male entrepreneurship running at twice the level of that for women. As Heide Inhetveen and Mathilde Schmitt show for Germany in *Chapter Five*, for women to move into self-employment, and make a success of it, they need to overcome cultural barriers. Significantly, radical change in agriculture (and, more broadly, in rural society) can be successfully produced by the actions of farm women, but this does necessitate transgressing cultural boundaries. That this is not easy for many to achieve is demonstrated by Christina Scholten in *Chapter Six*. Drawing on an analysis of Swedish programmes to encourage entrepreneurship in rural areas, this research shows that women tend not to be considered 'businesswomen', according to the norms and values of the male-dominated contemporary business environment. The types of 'livelihood' business units that women seek establish tend to be frowned upon, with their focus on the service sector and their expected relatively small size

---

[1] Offering transparent messages on rural deprivation in this regard, the report of the Commission of the European Communities Directorate-General for Agriculture (2000, p.15) offers a variety of indicators of poor rural childcare services. Included amongst these are the messages that 60 per cent of children aged 3-6 in rural Denmark were in publicly funded childcare services in 1991 compared with 85 per cent in Copenhagen, that only 1 per cent of rural children under four were in day-care in the Netherlands in 1992, compared with 16 per cent in the main urban centres, and that, in Minorca, a 1990 survey found that no rural two year olds, only 10 per cent of three year olds and just 20 per cent of four year olds benefited from pre-school services, compared with figures of 50 per cent, 90 per cent and 100 per cent, respectively, in urban areas.

being regarded as contrary to prevailing emphases in industrial sectors that do not see businesses as complementary to household economies. Significantly, here, as with elsewhere, such as with 88 per cent of those who would like to start a business in Haugen's (2004, p.79) Norwegian study, women who wish to start a new business recognize that they needed advice and information on other women's experiences. Yet the approach of advisors is hardly suited to the prior experiences and hopes of women, for: 'What the advisors want, is that women, starting a business after long-term unemployment, have a business from which they can receive a salary from the first day' (Scholten, 2004, p.113). Comparing the Haugen and Scholten chapters here points to issues that require further exploration. For while Scholten found a pattern of bias against enterprises headed by women, which led some to take the view that their advice to those considering starting an enterprise was not to bother, Haugen cites research which shows that 60 per cent of Norwegian women who received installation grants to establish a business were still in operation five years later. With generally high failure rates for small businesses, this is a good survival ratio, which prompts the question of whether there are different survival rates (and establishment rates) across nations, as well as between rural and urban areas for businesses run by women (and why this is the case).

Of course, an explanation for such divergence is likely to involve more than simply rural-urban or cross-national differences, with bases for differentiation being potentially affected by place of residence within rural and urban zones, as well as by social class, ethnicity, age, sexual orientation and family support structures. Quite evidently, for example, evidence shows that it is younger women who are more likely to leave rural areas in order to seek wider career and social opportunities (e.g. Demossier, 2004). This is not simply as a result of a greater attachment to a 'home place' amongst those who are older, but also arises from the encouragement of parents. As Bettina van Hoven (2004, p.138) indicates in her study of women in difficult employment circumstances in the länder of eastern Germany: '... they had no intention of leaving their home village and seeking their fortune in the West. They felt emotionally bound to their homes in Uecker Randow but also admitted that they encouraged young people, even their own children, to leave and improve their skills and employability in the West' (see also Gourdomichalis, 1991). The role of mother-sibling relationships is too little understood in rural literature, yet there are tantalizing hints at a powerful bond that has the potential to bring significant change to rural areas. Mauleón (2004, p.41) offers a strong message in this regard in noting how the prospects of that would-be male dairy farm successors in the Spanish Basque country will view working conditions on the farm positively is strongly affected by the position of their mothers:

> It has become clear that the role played by mothers over 49 years old is crucial, because of their self-sacrifice both in supporting the offspring's decision to take over the farm, and in maintaining productive activity where the farmer only works part-time. These mothers, rather than women in general, seem to be the key aspect in family farming.

There are considerable similarities between the tone of this message and the results obtained by Robina Mohammad (2002) in her analysis of mothers and daughters in the southern Spanish city of Málaga. Whether or not the support afforded by mothers in these studies is linked to particular regional or national cultural differences or links to other socio-cultural processes and resources is difficult to assess, given the relative absence of research on mothers and their adult children within the literature.

Personalities and Personal Resources

The interactions of different 'forces' that bear on women's lives in rural areas is brought out in many studies on rural employment. In Haugen's (2004) work, for example, she shows how the restrictions of weak local job opportunities have a particularly detrimental effect on women seeking paid employment in rural Norway. Here it was found that a group of women existed in rural areas who did want an off-farm job, but for whom a lack of relevant qualifications (whether vocational or appropriate labour market experience) interacted with a lack of suitable jobs, to thwart them in the realization of their aspirations. This combination also provided a focal point in Bettina van Hoven's investigation in *Chapter Seven* on rural women in the former East Germany. Reminiscent of Brettell's (1986) *Men Who Migrate, Women Who Wait*, this chapter explores how women have been unable to tap into job opportunities that are frequently a long distance from home, with increasing gender stereotyping, a withdrawal into the private sphere, exclusion from politics at large, the massive loss of jobs, a lack of worker skills, poor transport and a lack of business-oriented initiatives in a formerly state-dominated employment setting, all leading to a position where those women who are over 50 are taking on the appearance of a 'lost generation'. What such messages bring out is the manner in which, as Haugen (2004) argues, education and training programmes need to be adapted to local labour needs.

That training (for the workplace in particular) is commonly not being provided in rural areas in a manner that is conducive to furthering gender equality is already recognized within the literature (e.g. Shortall, 1996; Bock, 2004). There is little doubt that this needs addressing, with women often being poorly represented amongst those who attend training courses in fields that are linked to prevalent job availability in local economies. Thus, in rural Greece, women make up less than 20 per cent of trainees for subjects like animal husbandry, arboriculture, horticulture and agricultural machinery (Commission of the European Communities Directorate-General for Agriculture, 2000, p.12). Exploring one initiative that has sought to change this situation, by offering courses that are directed towards the needs of women, as well as being delivered in ways that should enhance prospects of achieving relevant training, even in 'peripheral' rural regions, in *Chapter Eight* Ingunn Limstrand and Marit Stemland examine how the Northern Feminist University in Steigen, Norway, has enriched women's lives.

Geographical Location

In their account of education initiatives in the periphery, Limstrand and Stemland emphasize that improvements in communications technology have removed many previous handicaps to rural women's self-advancement through training. This is a point that has been made by other commentators, as with the Commission of the European Communities Directorate-General for Agriculture (2000, p.12), which picks out a trend that is growing within Europe, here using Spain as an example:

> In Spain, young rural people generally have a lower level of training and education than their urban counterparts, although the difference is becoming less marked. Among 20-24 year olds, few gender differences exist, and education is not an explanation of women's marginalization in the rural economy.

Exploration of the (early) life course of those who live in rural areas similarly points to little rural disadvantage, nor of noteworthy gender differences between those who were brought up in rural or urban environments (e.g. Bynner, 2000), although this message comes with the caveat that many who start in rural zones leave for career reasons, with some evidence that many young people want to leave rural areas;[2] although it seems that many who think about leaving do not manage to do so (e.g. Ford *et al.*, 1997). But any seeming absence of rural-urban differentiation does not mean that geographical location is unimportant.

One of the key reasons for a diminishing of rural-urban contrasts is migration, whether from rural areas or into rural areas, but this does not discredit the possibility that geographical differences are not bound into the selective nature of migration streams. It is clear, for example, that the absence of jobs and difficulties in securing housing in areas of high in-migrant demand, encourage young people to leave rural areas that lie outside easy commuting distances from major employment centres (e.g. Rugg and Jones, 1999; Lindsay *et al.*, 2003). Yet, according to some commentators, it is the absence of job opportunities that encourages families with more traditional values to seek solace in a rural home, with wives not expected to work (Murdoch, 1995). Of course, such visions contradict growth tendencies toward a noteworthy increase in women's engagements with paid-work in rural areas (Bock, 2004), but such ideas do point to possible associations between geographical location and differences in rural lifestyles. Here the chapters in this book have clear messages about differentiation, which draw attention to Little's (2002) stress that there is wide variation in employment structures in rural areas. As Bock (2004) emphasizes in this regard, rurality does not come across as a fundamental constraint in itself, but it can reinforce the impact of other dimensions of distinction, like local labour market conditions, cultural predispositions and personal resources (see also Little and Panelli, 2003). As Sally Shortall (2003, p.5) has put it:

---

[2] For example, in a multi-site British rural study, DTZ Pieda Consulting (1998, p.132) found that only 31 per cent wished to stay in the local area.

It would be nonsensical to suggest that rural and urban women do not face many of the same issues. Educational qualifications, resources, social class and other social factors strongly influence the position of women in society, regardless of whether they live in rural or urban areas. Nonetheless, location does matter. There are particular features of rural life that impact on women. Childcare provisions are a concern for all women, yet childcare provisions in rural areas pose particular questions because of population density and travel. The same is true of employment policies. Rural Development Programmes have a defined rural focus, and they may or may not impact on men and women differently. Farming is an occupation particular to rural areas, and it too holds particular implications for women.

It is in recognizing and articulating the interaction effects that exist between different conditions in rural areas and geographical location that a deeper understanding of space and place considerations in women's lives will be forthcoming. To make real progress in this sphere, comparative investigations of similar issues and processes in a variety of localities are needed, in order to tease out the potency of geographical effects.

Policy Frameworks

This point could also be made with regard to temporal change. Here the introduction of new policies provides a particular framework for analysis, since policy frameworks are liable to much shorter term fluctuation than cultural traditions, labour market contexts, personal resources and relative geographies. Understandably, all of these are inter-related, as Bock and Haan (2004) make clear, in noting how attitudes toward married women taking paid-work positions have altered in The Netherlands from condemnation in early post-1945 decades, as being harmful to family life and detrimental to children's health, to something to be encouraged in the 1980s in order to help modernize the economy. While van Hoven's (2004) chapter in this volume reminds us that local labour market conditions can be rapidly disrupted by changes in political orientation, the fact that policies are potentially subject to short-term change should focus our minds, for some of the conditions that are particular handicaps in rural areas are amenable to quick mitigation. Take the case of the labour market oriented AMO courses that help those in Norway who are registered as unemployed to train for new work (with women as 58 per cent of participants these courses). Owing to the rules for participation in these courses, women on farms, who are commonly not registered as unemployed, are not eligible to take part (Haugen, 2004). In similar vein, working unpaid on a farm, which many women have done, commonly disqualifies them from the full benefits of state assistance, including pensions. These are issues that could be adapted quickly by changes in regulations, such that the 'reality' of women's work situations could be appropriately acknowledged. After all, it is not as if governmental agencies are unaware of the 'realities' of women's work patterns and prospects in rural areas, as various official publications on this theme demonstrate (e.g. Commission of the European Communities Directorate-General for Agriculture, 2000).

This provides just one example of a bias that exists in the manner in which state policies are formulated so as to place (unintended or not) additional pressures on rural residents. Stated more generally, Limstrand and Stemland (2004, p.156) remind us that Lipton's (1977) urban bias hypothesis is not without relevance in the countrysides of Europe today:

> The political climate and central tendencies in state rural policy today are not envisioned from the perspective of these areas. Rural policy has become an 'in-spite-of' strategy, aimed principally at reducing the consequences of development. There is an important need for capacity-building, both in traditional production spheres and in potential new sectors.

This again draws attention to the inter-connectedness of frameworks that channel opportunities for women in rural areas. Capacity-building is not something that happens overnight but requires adjustments in vision and purpose, which can be eased or intimidated by the 'realities' people face. As Bennett (2004) has recently made clear, even when change in particular circumstances seems to herald new openings for changed circumstances, the persistence of old attitudes and behaviour norms, alongside the discomforts of change, can limit capacities or desires to grasp initiatives. Changes in policy frameworks can raise capacities for rural women in the short-term, but adjustment to only one of the frameworks that impinge on women's lives is not enough. Broader and deeper adjustments are required, that build on opportunities for making short-term improvements, while also confronting more stubborn norms and action paths that have so far been reluctant to change rapidly.

References

Baylina, M. and Bock, B.B. (2004) Gender and rural development in academic discourse, in H. Goverde, H.J. de Haan and M. Baylina (eds.) *Power and Gender in European Rural Development*, Ashgate, Aldershot, pp.95-98.

Bennett, K. (2004) A time for change? Patriarchy, the former coalfields and family farming, *Sociologia Ruralis*, 44, pp.147-166.

Berg, N.G. (2004) Discourses on rurality and gender in Norwegian rural studies, in H. Goverde, H.J. de Haan and M. Baylina (eds.) *Power and Gender in European Rural Development*, Ashgate, Aldershot, 127-144.

Bock, B.B. (2004) It still matters where you live: Rural women's employment throughout Europe, in H. Buller and K. Hoggart (eds.) *Women in the European Countryside*, Ashgate, Aldershot, pp.14-41.

Bock, B.B. and Haan, H.J. de (2004) Rural gender studies in The Netherlands, in H. Goverde, H.J. de Haan and M. Baylina (eds.) *Power and Gender in European Rural Development*, Ashgate, Aldershot, pp.106-126.

Brassloff, W. (1993) Employment and unemployment in Spain and Portugal: A contrast, *Journal of the Association for Contemporary Iberian Studies*, 6(1), pp.2-24.

Breeze, J., Fothergill, S. and Macmillan, R. (2000) *Matching New Deal to Rural Needs*, Countryside Agency Publication CAX 39, Wetherby.

Brettell, C.B. (1986) *Men Who Migrate, Women Who Wait: Population and History in a*

Portuguese Parish, Princeton University Press, Princeton, New Jersey.

Bynner, J., Joshi, H. and Tsatsas, M. (2000) *Obstacles and Opportunities on the Route to Adulthood: Evidence from Rural and Urban Britain*, The Smith Institute, London.

Clark, M.A. (2000) *Teleworking in the Countryside: Home-Based Working in the Information Society*, Ashgate, Aldershot.

Commission of the European Communities Directorate-General for Agriculture (2000) *Women Active in Rural Development*, Office for Official Publications of the European Communities, Luxembourg.

Commission of the European Communities Directorate-General for Agriculture (2002) *Newsletter 45*, Office for Official Publications of the European Communities, Luxembourg.

Demossier, M. (2004) Women in rural France: Mediators or agents of change?, in H. Buller and K. Hoggart (eds.) *Women in the European Countryside*, Ashgate, Aldershot, pp.42-58.

DTZ Pieda Consulting (1998) *The Nature of Demand for Housing in Rural Areas*, Report for the Department of the Environment, Transport and the Regions, Edinburgh.

Ford, J., Quilgars, D., Burrows, R. and Pleace, N. (1997) *Young People and Housing*, Rural Development Commission, Salisbury.

Gourdomichalis, A. (1991) Women and the reproduction of family farms: Change and continuity in the region of Thessaly, Greece, *Journal of Rural Studies*, 7, pp.57-62.

Hadjimichalis, C. (1994) The fringes of Europe and EU integration: A view from the South, *European Urban and Regional Studies*, 1, pp.19-30.

Hadjimichalis, C. (2003) Imagining rurality in the New Europe and dilemmas for spatial policy, *European Planning Studies*, 11, pp.103-113.

Halliday, J. (1997) Children's services and care: A rural view, *Geoforum*, 28, pp.103-119.

Haugen, M.S. (2004) Rural women's employment opportunities and constraints: The Norwegian case, in H. Buller and K. Hoggart (eds.) *Women in the European Countryside*, Ashgate, Aldershot, 59-82.

Hillebrand, H. and Blom, U. (1993) Young women on Dutch family farms, *Sociologia Ruralis*, 33, pp.178-189.

Hoven, van B. (2004) Rural women in the former GDR: A generation lost?, in H. Buller and K. Hoggart (eds.) *Women in the European Countryside*, Ashgate, Aldershot, pp.123-140.

Hughes, A. (1997) Rurality and 'cultures of womanhood', in P.J. Cloke and J. Little (eds.) *Contested Countryside Cultures*, Routledge, London, pp.123-137.

Inhetveen, H. and Schmitt, M. (2004) Feminization trends in agriculture: Theoretical remarks and empirical findings from Germany, in H. Buller and K. Hoggart (eds.) *Women in the European Countryside*, Ashgate, Aldershot, pp.83-102.

Jentsch, B. and Shucksmith, M. (2004, eds.) *Young People in Rural Areas of Europe*, Ashgate, Aldershot.

Limstrand, I. and Stemland, M. (2004) Can education be a strategy for developing rural areas?, in H. Buller and K. Hoggart (eds.) *Women in the European Countryside*, Ashgate, Aldershot, pp.141-159.

Lindsay, C., McCracken, M. and McQuaid, R.W. (2003) Unemployment duration and employability in remote rural labour markets, *Journal of Rural Studies*, 19, pp.187-200.

Lipton, M. (1977) *Why Poor People Stay Poor: A Study of Urban Bias in World Development*, Temple Smith, London.

Little, J. (1994) Gender relations and the rural labour process, in S.J. Whatmore, T.K. Marsden and P.D. Lowe (eds.) *Gender and Rurality*, David Fulton, London, pp.11-30.

Little, J. (1997) Constructions of rural women's voluntary work, *Gender, Place and Culture*, 4, pp.197-209.

Little, J. (2002) *Gender and Rural Geography*, Prentice-Hall, Harlow.

Little, J. and Panelli, R. (2003) Gender research in rural geography, *Gender, Place and Culture*, 10, pp.281-289.

Mauléon, J.R. (2004) Family strategies and farming changes: The case of family farming in the Basque Country, in H. Goverde, H.J. de Haan and M. Baylina (eds.) *Power and Gender in European Rural Development*, Ashgate, Aldershot, pp.32-43.

Mohammad, R. (2002) *The State, Governance and Visions of Womenhood in Spain's Voyage to the Centre*, PhD thesis, King's College London.

Morin, E. (1970) *The Red and the White: Report from a French Village*, Random House, New York.

Morris, C. and Evans, N.J. (2001) 'Cheese makers are always women': gendered representations of farm life in the agricultural press, *Gender, Place and Culture*, 8, pp.375-390.

Murdoch, J. (1995) Middle class territory? Some remarks on the use of class analysis in rural studies, *Environment and Planning*, A27, pp.1213-1230.

Navarro, C.J. (1999) Women and social mobility in rural Spain, *Sociologia Ruralis*, 39, pp.222-235.

Persson, L.O. and Westholm, E. (1993) Turmoil in welfare system reshapes rural Sweden, *Journal of Rural Studies*, 9, pp.397-404.

Philip, L.J. and Shucksmith, M. (2003) Conceptualizing social exclusion in rural Britain, *European Planning Studies*, 11, pp.461-480.

Rugg, J. and Jones, A. (1999) *Getting a Job, Finding a Home: Rural Youth in Transition*, Policy Press, Bristol.

Scholten, C. (2004) Partnerships for regional development and the question of gender equality, in H. Buller and K. Hoggart (eds.) *Women in the European Countryside*, Ashgate, Aldershot, pp.103-122.

Shortall, S. (1996) Training to be farmers or wives? Agricultural training for women in Northern Ireland, *Sociologia Ruralis*, 36, pp.269-285.

Shortall, S. (2003) *Women in Rural Areas: A Policy Discussion Document*, The Rural Community Network Northern Ireland, Cookstown.

Chapter 2

It Still Matters Where You Live: Rural Women's Employment Throughout Europe

Bettina Bock

Introduction

Mary Braithwaite (1994) inventoried the employment situation of rural women at the end of the 1980s and at the start of the 1990s in the 12 countries constituting the European Union at that time. She concluded that throughout Europe women in rural areas had less employment opportunities compared both to men in rural areas and women in urban areas. In the meantime a lot has changed. There are now 15 countries in the European Union and Europe is on the verge of another significant enlargement. It is time to take a new look at women's employment in rural Europe, and to discover whether the employment situation of rural women has changed and, if so, in what respects.

This chapter starts its examination of these issues by comparing national and regional employment statistics. But in order to appreciate fully the enormous variation that exists in rural women's employment situation more detailed studies are needed. Consequently, in the second section of the chapter, local case studies are examined, so as to understand the consequences of rural women's continuing unemployment and under-employment. But to explain rural women's employment position, its variation across regions and its development over time, comparative studies are of outmost importance. Unfortunately, such studies are still rare, as most research on rural women's employment consists of small-scale local studies.[1] By pointing out the factors influencing rural women's employment and explaining their interaction, this chapter puts forward an analytic framework that might guide comparative research and add to the development of theoretical insight. This framework is presented in the third section of the chapter. Comparison of regions using the framework suggested in this section should make it possible to explain differences in employment, not only between men and women but also amongst

---

[1] Sometimes local case studies use a comparative framework, as with the work of Overbeek and associates (1998).

rural women. This in turn might add to the development of effective, tailor-made policies that improve rural women's employment, which is the focus of the commentary offered in the final section of the chapter.

Employment and work are often used as interchangeable concepts, resulting in work being measured by engagement in formal employment. As feminist researchers have pointed out repeatedly, a large part of women's work remains hidden in this way (Luxton, 1997). This is because such measures of 'work' commonly disregard, among others, unpaid (family) labour and work in the grey and black labour markets (Armas, 1999; Plantenga and Rubery, 1999). Applying a limited definition of work not only underplays what women do but also reconstructs, time and again, an image of women as non-workers and 'inactive' citizens. This obstructs our insight on women's labour engagement. But, while recognizing this limitation, this chapter focuses on rural women's formal employment and, thus, on rural women's engagement in regular (paid) jobs in the formal labour market.[2] There are pragmatic reasons for adopting this approach. Most obvious in this regard is that, in order to be able to make use of (cross-national and intra-national) statistics, a formal definition has to be applied. Moreover, the main ambition of this chapter is not to measure the amount of work women and men are doing but to point to the factors that may explain the (in)accessibility of paid-work and formal labour markets in rural areas. The invisible and informal workload of rural women is not forgotten in this chapter but is interpreted in the first place as one of the restrictions that hamper women's access to formal (paid) employment.

Of course, if the aim is to explore the paid-work position of rural women, some clarity is needed on the interpretation of rural areas that is to be used. 'Rural' is a highly contested concept, that evokes a lot of discussion about how rural areas should be defined. This chapter consciously refrains from participating in this debate, as defining 'the rural' is not among its objectives. Quite the contrary, its aim is to get an overview of the diverse opportunities offered to women living in rural areas. Therefore this chapter adopts the geographical definition of rural that is used in most employment statistics. This definition distinguishes between urban, intermediate and rural areas on the basis of population density.

The Employment Situation of Rural Women Throughout Europe

It is difficult to give a picture of the average employment situation of rural women, as employment prospects differ enormously by age, marital status, family stage and household situation, as well as by ethnic group and nationality, by education and

---

[2] Available statistics measure rural women's participation in paid-work, but this engagement might be located in either an urban or a rural labour market. The focus in this chapter is, therefore, on the ability of women and men living in rural areas to find employment anywhere. This chapter does *not* concentrate on the quality of rural labour markets. Of course, the more remote rural areas are, the more women and men are tied into local labour markets.

by work experience. Research has shown time and again that all these variables affect women's chances of employment. To give just one example, a growing number of rural women have a university degree, but even today lots of rural women have not finished primary school and are unable to read or write (Overbeek *et al.*, 1998; Armas, 1999). Another distinction of note concerns the employment position women have or aspire to. Rural women engage in a wide variety of forms of paid-work. Some have regular, temporary and flexible jobs, amongst which women are entrepreneurs, family workers, seasonal workers, home workers, illegal workers, unpaid workers and volunteers. Moreover, the employment situation of women varies considerably between and within countries. Statistics demonstrate the latter very clearly but do not allow for in-depth analyses of particular groups of women in specific regions. What Mary Braithwaite complained about in 1994 is still true today. There is still a lack of statistical material on the employment of rural women, and there is an absence of detailed knowledge about employment resources and the obstacles women face regarding paid-work involvements in the various areas of the European Union, and especially in the countries that are soon to join the Union (FAO, 1996; de Rooij and Bock, forthcoming).

Moreover, what statistics are available generally have serious limitations as regards women's work, for they focus on formal employment. Eurostat statistics offer an important advantage in this regard, as Eurostat adopts a broad definition of employment. This definition holds that persons in employment are all those who undertook any work for pay or profit of at least one hour in duration during the reference week that is used for the Labour Force Survey.[3] Applying this definition should result in the registration of a considerable part of women's unofficial and unpaid work (including family labour), presupposing of course that women report on these activities. The definition of unemployment is more problematic as it still does not take into account those women who would like to work, or work more hours, but do not see the point of seeking work and registering as unemployed. Consequently, statistics still undervalue the extent of women's unemployment and under-employment (Braithwaite, 1994; Commission of the European Communities, 2000b). Keeping this in mind, statistics are nonetheless helpful in giving a quick overview of differences between men and women, between countries and regions, and on how employment contexts have changed over time.

---

[3] For Eurostat, the employment rate refers to persons in paid-work aged 15-64 years as a percentage of total population of this age. Persons in employment are those who, during the week data are collected for the Labour Force Survey, undertook work for pay or profit for at least one hour. Persons who are not working but have jobs from which they were temporarily absent count as employed. Family workers are included as employed. The unemployment rate is the percentage of the total economically active population that is unemployed. The economically active population (or labour force) is the sum of employed (as employees or as self-employed workers) and unemployed. According to the criteria of the International Labour Organization (ILO), the unemployed are those aged 15 and over who (i) are without paid-work as an employee or in self-employment, (ii) are available to start work within the next two weeks, and (iii) have sought employment during the previous four weeks (see (un)employment statistics at http://www.europea.eu.int/comm/eurostat/public/datashop).

Table 2.1 European national average employment figures for men and women

	Employment						Unemployment		
	% women				% men		% women		% men
	1991	1996	1999	2002	1999	2002	1998	2002	2002
EU	50*	50	52	56	72	73	12	9	7
Austria		58	59	61	77	75	6	5	5
Belgium	43	45	50	51	68	68	12	8	6
Denmark	70	50	71	73	81	80	7	4	4
Finland	68	60	64	67	70	71	13	10	10
France	51	52	54	56	68	70	14	10	8
Germany	57*	55	57	59	72	72	10	8	9
Greece	35	39	41	43	74	72	17	15	6
Ireland	36	43	52	55	74	75	8	4	5
Italy		36	38	42	67	69	17	13	7
Luxembourg	44	44	49	52	75	76	4	4	2
Netherlands	49	55	62	66	81	83	6	3	3
Portugal	57	55	60	61	76	76	6	5	5
Spain	31	32	38	44	68	73	27	16	8
Sweden	78	69	70	73	74	76	8	5	5
UK	61	63	64	65	78	78	5	4	4
	1991	1997	1999	2002	1999	2002	1998	2002	2002
Accession states			55	50	67	62	16	15	14
Bulgaria				48		54	16	17	19
Cyprus			50	59	79	78	4	4	3
Czech Rep.	48	60	57	57	74	74	8	9	6
Estonia	47	61	58	58	66	66	9	9	10
Hungary	42	45	49	50	62	63	8	5	6
Latvia	49		54	58	65	64	14	12	15
Lithuania	48		61	57	69	64	11	13	13
Poland	45	52	52	47*	64	57*	12	21*	19*
Romania		61	60	53**	70	65**	6	8**	9**
Slovakia	46		52	51	64	62	15	19	19
Slovenia	47	58	58	59	67	69	8	6	6

Note: Employment is defined as participation in paid-work. There are no figures for Malta, which is excluded from the list of accession countries.
* Eurostat estimation. ** These figures are not comparable with earlier years as methods and definitions changed.

Sources: Employment rates for EU 1991-1999 and CEEC countries 1997-1999 are taken from Commission of the European Communities (2001a); EU and CEEC unemployment rates for 1998 are from Weise *et al.* (2001); employment and unemployment rates for 2002 are from Franco and Jouhette (2003) for EU countries and from Franco and Blöndal (2003) for CEEC countries; CEEC figures for 1991 and AC average figures are from FAO (1996).

Providing a sense of such differences, Table 2.1 describes the employment situation of women throughout Europe, with considerable differences between northern, western, southern and eastern European countries readily apparent. What Table 2.1 shows is that, in northern countries, female employment is generally high, with unemployment rates nearly equal for men and women. In western Europe, women's employment is rising but its level is still moderate. Here unemployment amongst women is decreasing but it is still somewhat higher for women than for men. In southern Europe differences between men and women are at their most impressive. Here female employment levels are at their lowest and female unemployment is high compared with other countries. However, the general trend in the south is the same as elsewhere, with women's employment rising and unemployment falling, at least in some countries. As for the eastern European (and new accession) countries, here the situation resembles northern and western Member States of the European Union, in so far as the situation in 1999 and 2002 is concerned. In eastern Europe relatively few women are employed compared to men, with unemployment levels rather high for both of them. Since 1998 female unemployment has risen or remained stable in all countries, except for Hungary, Latvia and Slovenia, where female unemployment has decreased somewhat (Behrens, 2002a).

As explained already, the Eurostat statistics in Table 2.1 show men and women as employed if they work for pay or profit for at least one hour during the reference week. As a result, labourers and employees, entrepreneurs and family-workers, and full-time and part-time workers, are all accounted for. However, what Table 2.1 conceals is the gender-specific character of employment categories. Throughout Europe, it is almost exclusively women who work as family labourers, whereas men are overly represented amongst entrepreneurs (Franco and Winqvist, 2002b). It is again men who dominate full-time work categories, whereas almost all part-time workers are women (Franco and Winqvist, 2002a).

For the spatial distribution of family labour and for part-time work, statistics are incomplete as well. Family labour is present in all countries and is prominent in urban as well as rural areas, but part-time work is rather unequally distributed across countries. It is most prominent in western Europe, especially in the Netherlands, but it is comparatively rare in southern and eastern Europe (Olsson, 2000; Commission of the European Communities, 2001c). Yet statistics on the regional predominance of part-time work are lacking. The different incidence of part-time work across countries can be explained by differences in labour market characteristics, by union politics, by average wage levels and by the availability of mechanisms that enable a reconciliation of work and family duties. In a number of West European countries it is part-time work that has been the most important means by which women reconcile paid-work and family. This is especially true for Belgium, Germany, Luxembourg, the Netherlands and the UK (Franco and Winqvist, 2002a). Women in northern Europe have other provisions at their disposal, like sufficient childcare institutions and long periods of maternity leave, which make it easier to accept full-time jobs. The same used to be true in many of the countries that will soon join the Union, but most of these provisions have been abolished or have become unaffordable in the last decade. In western Europe, relatively high wages (and the concept of a family wage) allow families to live on one (and a half) incomes. But, until recently, trade unions in several European countries, such as Finland, Germany and Italy, strongly opposed the legal introduction of part-time employment, as it was argued that this would weaken the position of employed workers (Pfau-Effinger 1994; den Dulk, 2001, 2002). By contrast, in countries like the Netherlands, trade unions have fought (successfully) for equal rights and treatment for full-time and part-time employees. Yet proponents and opponents of part-time work have both been right over their expectations. Part-time work has enabled the percentage of working mothers to increase enormously over the years. But, at the same time, part-time work is the

Table 2.2 Women's unemployment in 1999 by degree of urbanization

European 'region'		Female unemployment rate			Total unemployment rate		
		Urban	Inter-mediate	Rural	Urban	Inter-mediate	Rural
	EU	11	10	12	10	8	10
Northern	Denmark	6	5	7	6	5	5
	Finland	9	14	14	9	13	13
	Sweden	5	6	8	6	6	8
Middle	Belgium	12	8	10	11	6	7
	Germany	9	8	12	10	7	10
	France	14	14	14	13	12	11
	Luxembourg	4	3	3	3	2	2
	Netherlands	4	6	6	4	4	4
	Austria	6	5	4	6	4	4
	Ireland	5		6	6		6
	UK	6	4	5	7	5	6
Southern	Greece	20	17	12	14	10	7
	Italy	17	15	18	13	10	12
	Spain	22	25	24	14	10	7
	Portugal	6	4	6	6	3	4
Accession	Hungary*	11		11	13		19

Note: * These figures are for 1994 (for villages versus towns).
Source: Commission of the European Communities (2001b).

most important reason for the huge and persisting income gap between men and women, as well as the over-representation of women amongst low-waged employees (Eurostat, 2000; Clarke, 2001).[4]

But so far only national employment statistics have been discussed. Table 2.2 offers a first step toward a regional analysis of women's employment, which provides for a differentiation of regions within countries on the basis of their degree of urbanization (based on population density).[5] At first sight differences in employment between men and women appear to be small in this table. It is only after further differentiation and comparison is made within the category 'rural regions' that significant variation in rural women's employment situation comes to the fore. Thus Table 2.3 and Table 2.4 compare predominantly rural regions with high and low female unemployment rates in member and accession states of the European Union. These tables clearly show that the employment situation of women is worse in the most remote rural areas of the European Union; which is a conclusion that reaffirms the work of other researchers (Braithwaite, 1996; Overbeek *et al.*, 1998; Terluin and Post, 2000). These remote rural areas are characterized by low population density, with agriculture having a dominant position in shares of total employment.

The higher chance of unemployment in these areas is due to various factors. First of all, the changing global economy negatively affects traditional rural employment sectors. Not only in agriculture, but also in the fishery and forestry sectors, it is becoming more and more difficult to make a living, either in waged labour or as entrepreneurs (Post and Terluin, 1997; von Meyer, 1997). As most work in these sectors has been seen traditionally as men's work (Eurostat, 2002), it is mainly men's employment that is threatened by these changes (de Rooij and Bock, forthcoming). But as these changes occur, new economic sectors gain importance, especially in services and ICT, and these lead to new opportunities for (self)employment (Bryden, 1997). Some rural regions, for instance Scotland and Ireland, have prospered as a result of changes in their rural economies, with both leaning heavily on new service and high technology sectors (Rural Development Committee, 2001). Indeed, it is often women who are capturing these new jobs, as their professional competencies are generally more in line with the demands of new jobs than those of men. But there is also evidence that men are opposed to taking up these jobs as they perceive working in services as not matching their male identity (Dianne Looker, 1997). However, in many, and especially in the

---

[4] Job-segregation is also important in explaining income-disparities, with women over-represented in the service sector and in lower paid occupations. But even taking differences in education and job type into account, an average income gap of 15% exists between men and women in the European Union, indicating that gender discrimination still plays an important role (Benassi, 1999).

[5] Urban areas are defined as having a density of 500 or more inhabitants per square kilometre, where the settlement population is at least 50 000 inhabitants. Intermediate areas have a density of 100 or more inhabitants per square kilometre, and are either situated adjacent to densely populated areas or have a population of at least 50 000 inhabitants. Rural areas are thinly populated zones that meet neither of these criteria (information given by Eurostat on request).

Table 2.3 EU rural regions with high and low female unemployment rates, by population density and dominant economic sector, 2001

	Female unemployment rate		Total unemployment rate		Popul-ation density	Dominant sector
	1998	2001	1998	2001		
Areas with low female unemployment rates						
Niederbayern (D)	5	4	5	4	109	Manufacturing
Ionia Nisia (GR)	6	7	4	7	87	Agriculture
Baleares (E)	17	9	12	7	146	Services
Franche Comté (F)	11	7	9	5	69	Manufacturing
S-E Ireland	7	3	8	3	52	Agriculture
Trentino-Alto Adige (I)	5	4	4	3	68	Services
Oberösterreich (A)	4	4	3	4	115	Agriculture
Centro (P)	3	5	3	4	72	Agriculture
Aland (FIN)	3	1	3	1	17	Services
Smaland (SE)	7	4	7	4	24	Services
Areas with high female unemployment rates						
Dessau (D)	26	20	22	17	134	Manufacturing
Sterea Ellada (GR)	22	22	13	13	43	Agriculture
Andalucia (E)	40	32	30	22	82	Agriculture
Languedoc-Rousillon (F)	20	17	18	14	83	Services
Sydsverige (SE)	10	7	10	6	91	Manufacturing
Calabria (I)	37	36	27	25	137	Agriculture
Tirol (A)	7	6	6	4	52	Services
Alentejo (P)	13	8	9	6	19	Agriculture
Itae-Suomi (FIN)	16	14	17	14	10	Agriculture
Border-Midland-W Ireland	8	5	9	5	52	Agriculture

Note: For this analysis, it is assumed that regions are predominantly rural if their population density is less than 150 inhabitants a square kilometre. The abbreviations for countries in this table are: A – Austria, D – Germany, E – Spain, F -France, FIN – Finland, GR – Greece, I – Italy, P – Portugal and SE – Sweden.

Source: Figures for 2001 from Behrens (2002b). Figures for 1998 from Behrens (2000b). The categorization of economic sectors is taken from Weise *et al.* (2001).

most remote rural regions, lost opportunities in traditional rural sectors are not being replaced by new opportunities (Lindsay *et al.*, 2003). In these regions it becomes more and more difficult for local people to make a living and (young) people start leaving the area in search for (self)employment elsewhere (Warren-Smith *et al.*, 2001). This in turn negatively affects the quality of life in these areas. With a smaller and less affluent population, various services and facilities become endangered, up to the point that schools and shops are unable to continue. The

Table 2.4 Rural regions with high and low female unemployment rates in accession countries, by population density and dominant economic sector, 2001

	Female unemployment rate		Total unemployment rate		Popul-ation density	Dominant sector
	1998	2001	1998	2001		
Areas with low female unemployment rates						
Sud-Vest (RO)	7	8	7	8	83	Agriculture
Západné (SK)	13	18	12	19	124	Agriculture
Mazowieckie (PL)	9	14	8	14	142	Manufacturing
Nyugat-Dunátú (HU)	7	3	7	3	89	Manufacturing
Jihozápad (CZ)	6	7	5	6	67	Agriculture
Areas with high female unemployment rates						
Východné (SK)	21	22	21	24	84	Agriculture
Warmínsko-Mazurskie (PL)	21	25	18	22	60	Agriculture
Eszak Magyar (HU)	12	9	14	11	96	Manufacturing
Yugoiztochen (BG)	21	23	19	23	64	Agriculture
Severozápad (CZ)	11	15	9	13	141	Manufacturing

Note: Regions with a population density below 150 persons per square kilometre are assumed to be predominantly rural. The abbreviations for countries in this table are: BG – Bulgaria, CZ – Czech Republic, HU – Hungary, RO – Romania, PL – Poland and SK – Slovakia.

Source: Figures for 2001 from Behrens (2002a). Figures for 1998 from Behrens (2000a). The categorization of economic sectors is taken from Weise *et al.* (2001).

tendency of most European governments to economize on public service provision is an important factor in this trend. As a result, life in remote areas becomes even more difficult and less attractive, especially for young people and young families. In regional employment statistics, the impact of such outmigration (or in-migration) is difficult to detect. Paradoxically, increased rates of outmigration by unemployed rural inhabitants improves regional unemployment figures, even though local employment opportunities are deteriorating. In the same way, the in-migration of new rural inhabitants may improve local employment statistics, even if these in-migrants commute long distances to work in urban centres, so no new local jobs have been created (Irmen, 1997).

The situation is similar in the accession countries. Many of the problematic areas in these countries have local labour markets dominated by agriculture, although (former) manufacturing belts have very serious employment problems as well (Table 2.4). With the de-collectivization of the agricultural sector many women lost their jobs and became unemployed because of the lack of alternative employment (FAO, 1996; Hoven, 2004). Some women and men started as farmers in their own right on land they received when agriculture was privatized and land was (re)distributed among labourers and former owners. In general, however, women seem to have regained less land than men (Bamberger, 1994). This is for two main reasons. First of all, before socialism few women had been formal owners of land and, thus, few were eligible to be re-constituted as owners after the turnover of the socialist system. Today, land is (re)registered in the name of the former owner, so property is most often listed as belonging to a man (Giovarelli and Duncan, 1999). Secondly, compared to men, fewer women held powerful managerial positions in agriculture during the socialist period, with holders of such positions proving to be very successful in influencing the distribution of land (and other resources) in their own favour during the transition period (Einhorn, 1995). The resulting lack of property rights for women creates extra disadvantages as it often determines access to credit, training programmes and production subsidies (UNDP, 1999; Fotev *et al.*, 2001; de Rooij and Bock, forthcoming). As a result, women, who started the transition period with less resources than men, might now have less chance of maintaining or improving their situation in the longer run (Burn and Oidov, 2001).

What should be added is the fact that the employment situation of rural women differs considerably within accession states. It is therefore important to take into account the character of national and regional labour markets when seeking to appreciate the employment situation of women. In Hungary, for instance, (rural) women's unemployment rates are lower than those of men. This is partly due to women's prominence in the tertiary sector, to the closing down of male-dominated heavy industrial plants and to women's relatively low wages (Emigh *et al.*, 1999). But lower unemployment is also caused by women's under-registration as unemployed. Thus, whereas men become unemployed when losing their jobs, many rural women in Hungary become 'economically inactive' (Morell, 1999). One key reason for this failure to register as unemployed is that women see few prospects of finding a new job. Unavailability of unemployment insurance is probably another reason in countries that do not provide state payments or only grant payments to unmarried men or women. Moreover, women's unpaid work is ever increasing. Under socialism the availability of services was designed to assist women to combine paid-work and family but such services have either vanished or become unaffordable now (Einhorn, 1993, 1995; Kligman, 1994). Moreover, many women have to cope on their own, as husbands have migrated to jobs in urban areas and other countries, leaving families behind (Kligman, 1994). Not having a paid job or earning enough money, many rural women put a lot of effort into home-plots for food for self-consumption and extra income (Bridger, 1997; Laas, 1999; Majerová, 1999; Meurs, 1999; Morell, 1999; Rangelova, 1999; Haukanes, 2001). Added to which, with a need to economize, many women have turned to the home-production and processing of food, as well as making clothes (Rueschemeyer, 1998; Burn and Oidov, 2001; Pine, 2002). So far it is unclear and highly questionable whether all this hard and time-consuming work is counted in employment statistics. There is some evidence of women successfully developing subsistence and informal economic activities into their own businesses (Momsen *et al.*, 1999; Verbole, 1999; Bak *et al.*, 2000). But, even in these cases, it is probable that their work is not counted unless their businesses are officially registered. At the same time, high workloads obstruct women's mobility and hence their ability to accept work at long distances from their homes.

Looking back at employment statistics for countries and regions and over time (e.g. Table 2.1), we may conclude that, on average, (rural) women's employment situation has improved in the last decade. Although there are significant gender differences in employment, rural women's paid-work has increased considerably in many European regions. Female unemployment decreased at the same time in most countries in the European Union (Table 2.3). In accession countries, however, female (and male) unemployment has generally increased (Table 2.4). Moreover, although differences between countries and regions have become less pronounced, they have not vanished. The employment situation of rural women in southern Europe is still more precarious, as it is in remoter rural regions, compared to more urbanized rural regions. There is also evidence that the economic and political transformation of central and eastern European countries has had detrimental effects on the social and economic position of rural women.

All that stated, we should recognize that, while statistical analysis helps provide an overview of the rural employment situation, to understand women's and men's employment situation more fully, and to explain their divergence, more detailed information on their employment and living conditions is necessary. Local studies are therefore of outmost importance. Such in-depth investigations can reveal interdependence between rural women's increasing employment and other changes in local societies. To give just a few examples, rural women's employment is influenced by a falling need for farm labour and by the increased necessity of off-farm income. Other factors impacting on changes in women's employment patterns include more women from non-farm backgrounds becoming the partners of male farmers and rising levels of education amongst women living in the countryside. Increased transport mobility resulting from improvements in infrastructure and the modernization of values associated with traditional rural ideologies on gender relations, have added to the pace of change (Blekesaune and Haugen, 1999; Olderup, 1999). But local studies also demonstrate that, despite an increasing number of rural women holding paid employment, there are still important gender differences in working conditions and the quality of work undertaken. For example, in many regions more women work with no contract or only with a temporary work contract, or they work as home workers or unpaid family workers (see Baylina and García Ramon, 1998; Armas, 1999). Rural women are also less successful in finding a job that requires the level of education they possess, even though young women are better educated than men in many regions (Dahlström, 1996; O'Hara, 1998; Momsen *et al.*, 1999; Rangelova, 1999; Ní Laoire, 2001). As a result, many rural women have jobs that do not require workers to have their level of education (Commission of the European Communities, 2000a). According to Moss and colleagues (2000), this is especially true for farm women.

Before turning to the specific constraints rural women encounter when looking for employment, and the factors explaining women's local employment situations, the implication of rural women's unemployment and under-employment are briefly touched on. The key issues here concern the risks involved for rural women, their families and rural societies as a whole, of unemployment and under-employment.

The Consequences of Rural Women's Precarious Employment Situation

Since the study of Mary Braithwaite (1994), it has become even more evident that lack of employment opportunities for women presents a serious problem for the vitality and viability of rural areas. Female outmigration from the countryside is increasing, especially from the more remote rural areas. This is true in Greece and Italy, in Ireland and Scandinavia (Dahlström, 1996; Gidarakou, 1999; Högbacka, 1999; Forsberg *et al.*, 2000; Ní Laoire, 2001) and in central and eastern European countries (FAO, 1996; Majerová, 1999; Rangelova, 1999). Not just in Greece, France and Spain (Gourdomichalis, 1991), but also in a densely populated country like the Netherlands, farm women complain that their sons are not able to find a wife for precisely this reason. Nobody knows how serious this problem really is, as there is hardly any research on the matter. It is a fact, however, that several marriage agencies specialize in the agricultural sector. These agencies organize excursions for farmers to countries in central and eastern Europe to find women with no prejudice against marrying a farmer. On the contrary, many women in these regions perceive marrying a farmer 'from the West' as a rescue from poor living conditions and an insecure future for themselves and their families. Even if such arrangements result in happy marriages, this can be considered a legal form of the trafficking of women, which can hardly be thought of as an acceptable solution for the future of rural areas.

Outmigration is a general phenomenon, especially amongst young people in remoter rural areas, but it is most frequent amongst young women, particularly better-educated young women (FAO, 1996; Kulcsar and Verbole, 1997; Commission of the European Communities, 2000a). In fact, several researchers have described how getting a better education is a consciously chosen strategy of young women and their mothers to escape the countryside (Gourdomichalis, 1991; Dahlström, 1996; O'Hara, 1998). In some regions it is mostly boys who stay behind, as they are more bound to farms and other local businesses (Moss *et al.*, 2000; Ní Laoire, 2001). In other regions, especially in central and eastern Europe, men tend to migrate to urban areas and other countries on a temporary or a permanent basis, leaving women, children and elderly people behind (Pine, 2002). Those men and women who are not able to find employment elsewhere often have enormous problems making ends meet. Although 'the transition to democracy' enhanced poverty in general, the situation seems most problematic in the remote rural areas of central and eastern Europe. This is caused by a lack of employment opportunities and, thus, income, but it is also a function of a loss of services, like schools, childcare, public transport, and medical services, that were guaranteed under socialism and are now non-existent or unaffordable (de Rooij and Bock, forthcoming).

Lack of employment opportunities is not the only reason young women move to cities. There is also the prospect of hard work, little free time, few options for entertainment and (often) a more traditional, conservative cultural climate which fits badly with the modern aspirations of rural women and girls (FAO, 1996; Giradakou, 1999; Olderup, 1999; Rangelova, 1999; Forsberg *et al.*, 2000; Safiliou, 2001; Haugen and Villa, 2003). Such women aim for a better life, having been fed by higher education and the media on the opportunities modernity offers women. This phenomenon is called the 'Ally McBeal effect' (Dahlström, 1996). Rural areas where female employment is most problematical are the areas most badly in need of stimulating actions and policies. Such policies are necessary because of the needs of women and because of the devastating effects of female depopulation.

Analysis of the Problems Rural Women Encounter and the Factors at Stake

Most research on rural women's employment consists of small-scale local or regional case studies. These are indispensable for gathering in-depth knowledge on factors that positively or negatively influence women's employment in specific rural areas. But to explain the diversity of rural women's employment and to develop a comparative gender-specific theory of rural employment, it is important to assemble, systematize and then compare this 'local' knowledge. This chapter tries to make a start with developing an overall theoretical and analytical framework, which may enhance insight on the diversity of rural women's employment situation and may be useful for setting up comparative research on this theme. In line with this aim, the model presented in Figure 2.1 distinguishes between structural, cultural and individual influences on rural women's paid employment, much as many general studies on women's employment present (e.g. van Doorne-Huiskes *et al.*, 1995). As these general studies deal mostly with urban women and do not take the difference between the urban and the rural into account, a geographical factor has to be added in order to explain the effect of 'rurality'. The differentiation between geographical, structural, cultural and individual factors is of course analytical, for in real life these factors are interrelated. Yet a differentiation is necessary to improve our understanding of the interaction and relative importance of factors that explain women's employment in specific localities. In what follows a brief discussion is provided on the content of structural, cultural, individual and geographical factors, taking into account what local studies have already revealed about their impact on rural women's employment position.

At the structural level of the regional economy, the character of the regional labour market and the availability of jobs, especially in the service sector, are important influences on women's paid-work (Dahlström, 1996). Research makes it perfectly clear that the cut-back in public spending and the concentration and reduction of rural services has detrimental effects on the employment situation of women (FAO, 1996; Högbacka, 1999; Commission of the European Communities, 2000b; Moss *et al.*, 2000). It is service industries in particular that attract (rural) women and seem to offer good employment opportunities (Overbeek *et al.*, 1998). These are the industries that have seen the most important growth levels in recent years, with ongoing growth in employment expected (Terluin and Post, 2000). Actual and future employment opportunities for rural women are thus probably best in areas where the service sector is growing. Moreover, the availability of part-time work, childcare and other services and arrangements are decisive, as

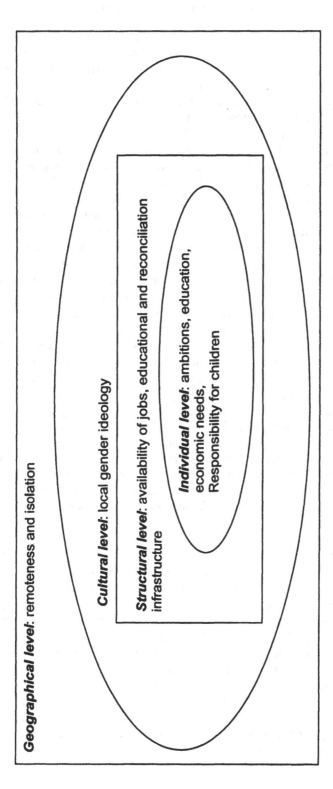

Geographical level: remoteness and isolation

Cultural level: local gender ideology

Structural level: availability of jobs, educational and reconciliation infrastructure

Individual level: ambitions, education, economic needs, Responsibility for children

Figure 2.1 Interaction of factors influencing women's employment at geographical, cultural, structural and individual levels

these help support the reconciliation of work and family (e.g. parental leave, emergency leave, flexible work hours; Braithwaite, 1996; Overbeek *et al.*, 1998; Halliday and Little, 2001). Education and training facilities are also important in sustaining the re-entrance of women into the labour market and the updating of their employability (Overbeek *et al.*, 1998). Last but not least, infrastructure and the availability of (public) transport are decisive for enabling women (and men) to accept paid labour outside their locality (Post and Terluin, 1997).

On a cultural level, local gender ideologies contain dominant norms and values concerning female employment and entrepreneurship, male breadwinning, working mothers and the division of household and family duties (Little, 1991, 1994; Bak *et al.*, 2000; Ogbor, 2000; Knudsen and Waerness, 2001; Mauthner *et al.*, 2001). Pfau-Effinger (1994, p.1370) summarized these issues under the concept of the 'family and integration model', which is '... the particular combination[s] of cultural norms and values in a society concerning the integration of women and men into different societal spheres, the gendered division of labour, and the societal sphere for caring for children and other dependents'. Pfau-Effinger (1994, 2000) uses this model in order to distinguish between nations and their 'gender cultures', but, following Sackmann and Häusermann (1994), a similar model may be used to explain regional differences in female employment behaviour. In both cases it is important for researchers to take the historical background of different models into account. Dependent on the specific circumstances of national and regional industrialization processes, specific behavioural patterns are stimulated and supported by norms and values defining what kind of childcare and what kind of female employment is considered appropriate. Such ideologies influence the behaviour of women, but also their identity, their ambition and assessments of the chances of succeeding, along with expectations on likely judgements by others. Even though women differ in their acceptance of dominant local norms and values, they cannot avoid being influenced by them. After all, the same norms and values affect the expectations and behaviour of significant others, among which are local employers, village councils, husbands and parents (and parents-in-law) (Little, 1991, 1994).

On an individual level, women's education and professional qualifications, along with their drive and ambition, are important factors influencing the resources women have at their disposal, and consequently impact on the extent to which women are able and ready to invest in finding a satisfying job (Overbeek *et al.*, 1998; Moss *et al.*, 2000). Having means of transport is of course an important resource as well, which is especially necessary to search for and then accept a paid job (Lindsay *et al.*, 2003). Overall, the interaction between aspiration, education and mobility is more effective in influencing women's employment behaviour than are local labour market characteristics. Even when a local labour market has little to offer those with high education qualifications and considerable ambition, this can stimulate women's willingness to increase their mobility and accept long distances between work and home (Overbeek *et al.*, 1998; Moss *et al.*, 2000). Furthermore, age and household situation are of great importance. The effect of age varies between countries. In some European countries, young women and men differ very little in the extent of their labour market participation. In 1997, for instance, there was no difference between the labour market participation of women and men in the age group of 15 to 29 years in Greece and men's participation rate was only slightly higher (2%) in Sweden (Commission of the European Communities, no date). In Finland, France, Ireland and the Netherlands only 6% more men compared to women had paid labour in that age group. In other EU countries, gender differences were remarkable even at this young age. This is true in Italy and in Denmark, where the participation rate of young men and women differed by 14% and 13%, respectively, in favour of young men.[6] In nearly all European countries women with small children leave the labour market, at least temporarily, or switch to part-time employment (Bock and van Doorne-Huiskes, 1995; Franco and Winqvist, 2002a; Tijdens, 2002). However, education and ambition do counteract this interrelationship, just as the support of a husband for his wife's career and his willingness to take his part in caring for children have an effect. Finally, the economic situation of the individual household is of great importance, as lack of income may force (rural) women to accept whatever work is available as part of the family's survival strategy (Perrons and Gonäs, 1998; Plantenga and Rubery, 1999; Kelly and Shortall, 2002).

The factors described above apply to the situation of women in general. To understand the situation of rural women, it is important to add another level of explanation, which encompasses their geographical context(s). These contexts are affected by population density and distance to urban centres; in short to the 'rurality' of an area. The more remote and geographically isolated an area is, in general the more difficult the situation is likely to be for women. This may be explained by a number of considerations.

For one, rurality affects local labour market structures. In many rural areas in Europe, agriculture is still the dominant employer, whereas the service sector, which is more important for female employment, has few jobs to offer. Long distances to employment concentrations, little public transport and bad roads restrict commuting to work (Irmen, 1997; Lindsay *et al.*, 2003). Moreover, husbands or fathers may already be migrating for work purposes, which renders a women's presence at home indispensable, in order to have someone to look after the family (Halliday and Little, 2001). Women may also be needed to take care of the farm, household food production and extra farm income generating activities. The latter is often found in eastern European (accession) countries but is also common in the middle of Italy (Bock, 1994) and Norway (Jervell, 1999). As men's engagement in paid-work promises a higher income, or men's chances of finding work are perceived to be better, it is more likely that male members of households will be the ones to commute or even migrate for work. This pattern is also more acceptable in many countries, as with (rural) Italy, where taking care of the home and the family is considered 'naturally' as a woman's job (Trifiletti, 1999).

The remoteness of rural areas might furthermore badly affect the resources of

---

[6] Gender-specific differences in labour market participation at a young age also reflects different enrolment rates in tertiary education. At the moment more girls than boys participate in tertiary education in most EU and CEEC countries (Dunne, 2001).

individual women and, thus, their ability to find satisfying employment. Their level of knowledge and experience may be low because there are few possibilities for training and education, or for the accumulation of work experience. Also the fact that channels for finding paid-work are often of an informal nature (e.g. Lindsay *et al.*, 2003), can work against women finding out about (non-advertised) job possibilities. As there is generally little difference in education between young men and women (Dianne Looker, 1997), lack of training is probably most constraining for older women. In former times rural women had little opportunity to obtain a professional education. Moreover, their education is most probably outdated today (e.g. Haugen, 2004). In this context, it is not surprising to find that many 'rural' mothers support their children, especially their daughters, in their efforts to get a good education, with the consequence that children are freed from helping in the household and on the farm as far as possible (Gourdomichalis, 1991; Bock, 1994; O'Hara, 1998; Ní Laoire, 2001).

Finally, it is important to recognize that the cultural climate in rural areas tends to be rather traditional and conservative concerning gender equality, especially in the most remote areas (FAO, 1996; Dianne Looker 1997). Thus, Little (1991, 1994) points out that the typical rural gender ideology is one of the most important constraints on rural women's engagement with paid-work. This ideology holds that rural women should be mothers and caretakers of the community first. This rural gender-ideology consequently stimulates and promotes women to stay at home, to take care of their family and household and to engage in voluntary work for the sake of the local community (e.g. Little, 1997b). This value disposition also influences local employers, convincing them to offer only temporary, low-paid and part-time jobs to women (Little, 1997a). As a result, women in rural areas not only miss encouragement and support to build a satisfying professional life but are often actively prevented from doing so.

The geographical character of rural areas or, in short, 'rurality' is not so much a constraint in itself but a reinforcing effect on other restrictions. It negatively affects the precarious balance of structural and cultural preconditions needed morally and practically to enable and empower individual women to make use of their professional capacities and aspirations and overcome the numerous obstacles they encounter. In many rural areas employment conditions are not as favourable as in cities to begin with, as the local economy and labour market are less diversified and reconciliation services, such as childcare provision, are lacking. Additionally, the cultural climate plays a more constraining role, as opportunities for women and men to escape from prejudice and gossip are more limited, with local social control more effective in rural than in urban areas (Haugen and Villa, 2003).

This interpretation resembles explanations of women's position by way of gender contract or gender arrangement theory (Pfau-Effinger, 1994, 2000; Duncan, 2000). In this theory gender-specific norms and values (or 'gender cultures') develop in close relation with structures and institutions that promote gender specific employment behaviour. Presupposing and re-enforcing each other, gender culture and structure form one coherent unity, the gender contract. As Duncan (2000) points out, gender contracts are spatially bound and differ not only between but also within countries. Forsberg and associates (2000) reveal how regional differences in female employment patterns within Sweden may be explained by regional variations in the national gender contract. The more equal regions are situated in the north; they are generally bigger in size and offer a lot of employment in the service sector. The more unequal regions are in the south, which are smaller in size with a labour market that is industry-based (Forsberg *et al.*, 2000). Research in Switzerland also reveals how regional gender contracts vary along language borders, influenced by ethnic, religious and political factors (Bühler, 1998, in Forsberg *et al.*, 2000). Regional variation in female employment patterns may thus be explained as the interaction of regional gender cultures and regional economic and institutional structures (Forsberg *et al.*, 2000). The resulting regional gender contracts determine women's roles, opportunities and expectations and, among others, women's employment patterns. This determination is, however, not absolute. First of all, women deviate from dominant gender contracts and act against societal expectations. Secondly, regional gender contracts may and do in fact change in the course of time.

Following Pfau-Effinger (2000) change is the result of discrepancies between expected and actual behaviour, time lags and variation within groups concerning beliefs and behaviour. Women and men are thus perceived as important actors in change. By orienting their behaviour according to innovative norms and values, they may introduce new gender practices and step-by-step change regional gender cultures and structures. Young rural women might form such a group, as long as they are not migrating from rural areas in search of better opportunities in regions (cities) that have already more equal gender contracts (Forsberg *et al.*, 2000; Ni Laoire, 2001; Demossier, 2004). Following Duncan (2000), culture is more important in determining gendered employment patterns than economic structures. Departing from that point, it might be expected that change starts from changing cultures and, thus, changing norms and expectations. But only longitudinal regional studies, embedded in comparative research, can reveal whether or not this is the case. Generally speaking, there is most probably an ongoing interaction between the cultural, structural and individual factors distinguished above.

A Comparison of Nordic Countries with those of Central and Eastern Europe

Differentiating between factors may guide the organization of comparative research on the employment situation of rural women by point to variation across research areas and populations that should be explored. But it may also help to structure and interpret already available evidence and explain why rural women have more employment problems in one region than another. To illustrate the usefulness of this analytical framework, we give two examples, one explaining the relative advantage of Nordic women, the other analyzing the specific situation in central and eastern Europe. Both examples are only briefly sketched, in order to serve as an illustration; a profound analysis of these respective regions of course requires a more thorough study than can be provided in this chapter.

The Nordic countries are often referred to as the vanguard of gender equality (den Dulk *et al.*, 1999; Rubery, 1999; Duncan, 2000; den Dulk, 2002). Amongst other reasons, this is due to the almost equal employment situation of men and women. This is not by coincidence, as examination of the structural and cultural preconditions and resources of individual women makes clear. Thus, the Nordic countries have a long tradition of government policy actively stimulating and supporting female employment by investing in facilities that help reconcile work and family engagements. Even in rural areas the government has created many jobs in the service-oriented public sector that offer rural women satisfying employment. With serious cut-backs in public spending this situation has changed in recent years, with resulting increased unemployment amongst rural women (Persson and Westholm, 1993; Safiliou, 2001). It is, however, typical of the Nordic countries that public opinion is strongly in favour of working mothers and gender equality in general. Last but not least education and aspiration levels among women and girls are generally high (Overbeek *et al.*, 1998; Blekesaune and Haugen, 1999; Högbacka, 1999; Jervell, 1999; Oldrup, 1999).

The accession states in central and eastern Europe (CEEC) present a specific case, where there is differentiation that yields its own specificities across structural, cultural, spatial and individual circumstances. The gender-contract of CEEC countries is special, as women were expected to engage in paid-work during the socialist period, so women's employment was normal and accepted (Kligman, 1994). On the other hand, much research has indicated that although female employment was nearly equal to men's (at least quantitatively if not qualitatively), gender-relations and gender-ideology were far from equal (Kligman, 1994; FAO, 1996; Momsen *et al.*, 1999; Morell, 1999; de Rooij and Bock, forthcoming).[7] Household and children remained a woman's responsibility. The reconciliation of work and family was guaranteed not by gender-equality but by public services and provisions (Einhorn, 1993, 1995; Kulcsar and Verbole, 1997; Laas, 1999; Morell, 1999). As these were lost as the political and economic system changed, CEEC women today miss the material and immaterial support necessary to combine work and family successfully (Kligman, 1994; Einhorn, 1995; FAO, 1996; Rueschemeyer, 1998; Majerová, 1999; Pine, 2002; de Rooij and Bock, forthcoming). As well as this, labour market structures changed fundamentally with the transition process, which resulted, amongst other things, in many women's (and men's) jobs being abolished. There are indications that it was especially in the more remote rural areas that this process hit hard (e.g. Fink *et al.*, 1994; Hoven, 2004), with these areas suffering from structural unemployment ever since. Women seem to have more problems in these areas finding new work compared to men because either family duties or a lack of educational resources limit their mobility. But again the situation differs across regions. Although in most regions men migrate to find work, in certain regions we find high levels of migration amongst women as well. In the gender contract of these regions, women and men are perceived as being equally responsible for maintaining the family economically (Pine, 2001).

---

[7] See Bridger (2001) for an analysis of changes in gender-ideologies in the Soviet Union before the transition.

Conclusion and Policy Relevance

Analysis of European statistics reveals that rural women's employment has increased overall in the last decade. Despite this, there is a significant difference in the employment conditions of men and women. On average men have better and safer jobs in terms of contracts, salary and correspondence between job requirements and educational levels. But differences between regions within Europe are huge and, as a result, differences between rural women demonstrate similar wide variation. In some rural areas employment opportunities are actually better for women than for men. In others, rural women have little chance of finding paid-work. Generally speaking, rural women's employment problems increase with the remoteness or 'rurality' of an area. To understand why this is the case and how differences between rural regions and between rural women may be explained, more comparative research on rural women's employment is necessary.

This chapter develops an analytical framework that may guide the design of comparative research and may serve as a starting point for developing a gender-specific theory of rural employment as well. On the basis of general (and mostly urban-oriented) research on women's employment, the factors that define the preconditions of women's employment are articulated on the structural, cultural and individual levels. In more concrete terms, the factors refer to (amongst others) the character of the local labour-market, local gender-specific norms and values, and women's individual resources, such as their education level and their ambition. To understand differences across rural areas, and between rural and urban areas, a geographical level of explanation has been added, that encompasses the influence of local context. By summarizing the results of already available research, this chapter takes the position that 'rurality' is not a constraint in itself but reinforces the effect of other structural, cultural and individual restrictions. Rurality negatively affects the precarious balance of structural and cultural conditions, which give women enough encouragement to invest in education and develop professional ambition, as well as to confront many existing 'moral' and practical objections to their involvement with paid-work. Research results so far indicate that rural (gender) culture and, thus, cultural factors are more effective employment constraints compared to structural constraints, such as lack of jobs or public transport. As has been shown above, ambition may overcome such 'logistical' problems. This prevalence of cultural factors is confirmed by other research that compares the female employment situation in different rural areas (Forsberg *et al.*, 2000). But it is also confirmed by research analyzing the general prosperity and employment situation in rural regions. This research has revealed that cultural factors, such as entrepreneurial tradition, regional identity and work ethic, are among the most important factors explaining the difference between lagging and leading regions (von Meyer, 1997).

The question remains then, to what extent is rural culture spatially bound and linked to geographically rural areas. In a recent article, Henderson and Hoggart (2003) seem to imply that rural cultures might exist also in urban areas and urban cultures might be present among rural inhabitants. Following their line of argument, rural culture and rural location should be treated as two separate concepts. Does this then mean that rurality as a geographical factor has no effect on women's employment after all? Following the thoughts developed before, rural location in itself offers indeed little explanation. It is in interaction with individual, cultural and structural factors that a rural location becomes a re-enforcer of constraints and difficulties, and this effect is stronger with increasing remoteness and spatial isolation. The strength of this interaction in different regions is something further research has to reveal. Research should also examine the interdependence of culture and location and clarify to what extent modern and traditional gender cultures differ between rural and urban areas. Departing from the gender contract perspective, culture and structure are necessarily interrelated because they build in continuous interaction. This interaction is spatially bounded. Culture and structure form a coherent unity but '… only under particular conditions of time and place' (Pfau-Effinger, 2000, p.268).

This analytical framework improves our insight on the interaction of multiple regional characteristics in restricting rural women's access to employment. It also enhances our ability to explain variation in rural women's employment situations. But approaching rural women's employment along these lines is also useful for developing (more) effective policy measures and instruments. So far researchers have been quite pessimistic about the effectiveness of rural policies when it comes to the improvement of rural women's employment circumstances (Braithwaite, 1994, 1996; Overbeek *et al.*, 1998; Overbeek, 1999; Oedl-Wieser, 1999). In this context, it is first noted that there are few gender-specific policy measures and instruments. Moreover, those policies that exist often take insufficient account of how various factors interact in the persistence of women's limited access to paid employment. Policies may, for instance, provide training courses with childcare facilities but without offering assistance in finding a (part-time) job afterwards. Whereas training followed by men is generally upgrading qualifications in promising sectors, a lot of the courses offered to women are introductory and have little effect on women's employability and employment chances (Oedl-Wieser, 1999). There is also evidence that women are under-represented in start-up incentives that encourage and support the unemployed to start their own business (Melis, 2003). In rural areas self-employment is not only widespread but is also an important means to create employment. For women it offers specific advantages to reconcile family duties and paid labour (Plantenga and Rubery, 1999). It is therefore important that start-up incentives are not only equally available for women but also embedded in a specific, tailor-made framework of support (e.g. Warren-Smith *et al.*, 2001).

In the long run, effective policies have to approach the local gender-system as a whole (Duncan, 1995), taking the multiplicity of structural, cultural, individual and geographical factors, plus the totality of rural women's working and living conditions into account (Oedl-Wieser, 1999). Effective policy also requires that policy-makers ensure that commitments to change are invested in over a long period, since change takes time. At the same time, as a number of researchers have pointed out, policy-makers need to involve women in the development of local policies in order to ensure that specific local conditions are taken into account sufficiently (e.g. Braithwaite, 1996). As research has demonstrated, rural women's participation in policy-making supports projects that are tuned to the needs and possibilities of local women. This certainly generates more interest amongst rural women and makes projects more accessible for them than most top-down projects are capable of achieving (Bock, 2002). Finally, research has made perfectly clear that, although gender-mainstreaming is an important and promising development, projects that are specifically women-oriented are still needed in order to guarantee the accessibility and effectiveness of employment projects for the most disadvantaged women (Braithwaite, 2000).

References

Armas, P. (1999) *Home-working in a peripheral region (Galicia, Spain)*, paper presented at a conference on 'Gender and Rural Transformations in Europe', 14-17 October 1999, Wageningen University.

Bak, M., Kulawczuk, P. and Szcesniak, A. (2000) *Providing Assistance to Women in Rural Poland: The Perspectives of Providers and Beneficiaries*, Institute for Human Sciences SOCO Project Paper 75, Vienna.

Bamberger, M. (1994) Gender and poverty in socialist economies: The World Bank experience, in N. Heyzer and G. Sen (eds.) *Gender, Economic Growth and Poverty*, International Books, Utrecht, pp.334-351.

Baylina, M. and García Ramon, M.D. (1998) Homeworking in rural Spain: A gender approach, *European Urban and Regional Studies*, 5, pp.55-64.

Behrens A. (2000a) Regional unemployment rates in the Central European candidate countries 1999, *Statistics in Focus – General Statistics 1*, Eurostat, Office for Official Publications of the European Communities, Luxembourg.

Behrens A. (2000b) Unemployment in the regions of the European Union 1999 *Statistics in Focus – General Statistics 3*, Eurostat, Office for Official Publications of the European Communities, Luxembourg.

Behrens, A. (2002a) Regional unemployment rates in the Central European Candidate Countries 2001, *Statistics in Focus - General Statistics 6*, Eurostat, Office for Official Publications of the European Communities, Luxembourg.

Behrens, A. (2002b) Unemployment in the regions of the European Union 2001, *Statistics in Focus - General Statistics 7*, Eurostat, Office for Official Publications of the European Communities, Luxembourg.

Benassi, M.P. (1999) Women's earnings in the EU: 28% less than men's, *Statistics in Focus - Population and Social Conditions 6*, Eurostat, Office for Official Publications of the European Communities, Luxembourg.

Blekesaune, A. and Haugen, M.S. (1999) *Time Allocation and Life Quality Among Norwegian Farm Women: Work On-Farm or Off-Farm - Does it Make Any Difference?*, Centre for Rural Research CRR-Paper 9.99, Trondheim.

Bock, B.B. (1994) Female farming in Umbrian agriculture, in L. van der Plas and M. Fonte (eds.) *Rural Gender Studies in Europe*, van Gorcum, Assen, pp.91-107.

Bock, B.B. (2002) *Gender, plattelandsontwikkeling en interactief beleid* [Gender, Rural Development and Interactive Policy-Making] PhD thesis, Wageningen University.

Bock, B.B. and Doorne Huiskes, A. van (1995) The careers of men and women: A life-course perspective, in A. van Doorne-Huiskes, J. van Hoof and E. Roelofs (eds.) *Women* and the European Labour Markets, Paul Chapman, London, pp.72-89.

Braithwaite, M. (1994) *Economische rol en positie van de vrouw op het platteland*, Bureau voor Publikaties der Europese Gemeenschappen, Luxembourg.

Braithwaite, M. (1996) Women, equal opportunities and rural development: Equal partners in development, *LEADER Magazine 5*. (http://www.rural-europe.aeidl.be/rural-en/biblio/women/art03.htm).

Braithwaite, M. (2000) *Mainstreaming gender in European Structural Funds*, paper for the workshop 'Mainstreaming Gender in European Public Policy', 14-15 October 2000, University of Wisconsin, Madison (http://eucenter.wisc.edu/Conferences/Gender/braith.htm).

Bridger, S. (1997) Rural women and the impact of economic change, in M. Buckley (ed.) *Post-Soviet Women: From the Baltic to Central Asia*, Cambridge University Press, Cambridge, pp.38-55.

Bridger, S. (2001) Image, reality and propaganda: Looking again at the Soviet legacy, in H. Haukanes (ed.) *Women After Communism: Ideal Images and Real Lives*, University of Bergen Centre for Women's and Gender Research, Bergen, pp.13-28.

Bryden J. (1997) Rural employment and the information highway, in R.D. Bollman and J.M. Bryden (eds.) *Rural Employment: An International Perspective*, CAB International, Wallingford, pp.447-459.

Bühler, E. (1998) Economy, state or culture? Explanations for the regional variations in gender inequality in Swiss employment, *European Urban and Regional Studies*, 5, pp.27-40.

Burn, N. and Oidov, O. (2001) *Women in Mongolia: Mapping Progress Under Transition*, United Nations Development for Women, Washington DC.

Clarke, S. (2001) Earnings of men and women in the EU: The gap narrowing but only slowly, *Statistics in Focus - Population and Social Conditions 5*, Eurostat, Office for Official Publications of the European Communities, Luxembourg.

Commission of the European Communities (no date) *The European Labour Market in the Light of Demographic Change*, Directorate of Employment and Social Affairs (http://europa.eu.int/comm/employment_social/soc-prot/labour_market/english.pdf).

Commission of the European Communities (2000a) *Women Active in Rural Development: Assuring the Future of Rural Europe*, Office for Official Publications of the European Communities, Luxembourg.

Commission of the European Communities (2000b) *Gender Equality in the European Union: Examples of Good Practices (1996-2000)*, Office for Official Publications of the European Communities, Luxembourg.

Commission of the European Communities (2001a) *Employment in Europe 2001. Recent trends and prospects*, Directorate General for Employment and Social Affairs, Brussels.

Commission of the European Communities (2001b) *Report from the Commission to the Council, the European Parliament, the Economic and Social Committee and the Committee of the Regions: Second Report on Economic and Social Cohesion*, Office for Official Publications of the European Communities, Luxembourg.

Dahlström, M. (1996) Young women in a male periphery – experiences from the Scandinavian north, *Journal of Rural Studies*, 12, pp.259-271.

Demossier, M. (2004) Women in rural France: Mediators or agents of change?, in H. Buller and K. Hoggart (eds.) *Women in the European Countryside*, Ashgate, Aldershot, pp.42-58.

Dianne Looker, E. (1997) Rural-urban differences in youth transition to adulthood, in R.D. Bollman and J.M. Bryden (eds.) *Rural Employment: An International Perspective*, CAB International, Wallingford, pp.85-98.

Doorne-Huiskes, A. van, Hoof, J. van and Roelofs, E. (1995, eds.) *Women and the*

European Labour Markets, Paul Chapman, London.

Dulk, L. den (2001) *Work-Family Arrangements in Organizations: A Cross-National Study in the Netherlands, Italy, the United Kingdom and Sweden*, Rozenberg Publishers, Amsterdam.

Dulk, L. den (2002) Werkgevers en de zorgende werknemer: Hoe Nederlandse, Italiaanse, Britse en Zweedse organisaties de combinatie arbeid en zorg faciliteren, *Gedrag en Organisatie*, 15(4), pp.225-239.

Dulk, L. den, Doorne-Huiskes, A. van and Schippers, J. (1999, eds.) *Work-Family Arrangements in Europe*, Thela-Thesis, Amsterdam.

Duncan, S.S. (1995) Theorizing European gender systems, *Journal of European Social Policy*, 5, pp.263-284.

Duncan, S.S. (2000) Introduction: Theorising comparative gender inequality, in S.S. Duncan and B. Pfau-Effinger (eds.) *Gender, Economy and Culture in the European Union*, Routledge, London, pp.1-24.

Dunne M. (2001) Women and men in tertiary education. *Statistics in Focus: Population and Social Conditions 18*, Eurostat, Office for Official Publications of the European Communities, Luxembourg.

Einhorn, B. (1993) *Cinderella Goes to Market: Citizenship, Gender and the Women's Movement in East Central Europe*, Verso, London.

Einhorn, B. (1995) Ironies of history: Citizenship issues in the new market economies of East Central Europe, in B. Einhorn and E. Janes Yeo (eds.) *Women and Market Societies: Crisis and Opportunity*, Edward Elgar, Aldershot, pp.217-233.

Emigh, R.J., Fodor, E. and Szelenyi, I. (1999) *The Racialization and Feminization of Poverty During the Market Transition in Central and Southern Europe*, European University Institute EU Working Paper RSC 99/10, Florence.

Eurostat (2000) Nearly one employee in seven in the EU is on low wages, *Eurostat News Release 94*, Office for Official Publications of the European Communities, Luxembourg.

Eurostat (2002) *The Life of Women and Men in Europe: A Statistical Portrait - Data 1980-2000*, Office for Official Publications of the European Communities, Luxembourg.

FAO (1996) *Overview of the Socio-Economic Position of Rural Women in Selected Central and Eastern European Countries: Bulgaria, Croatia, the Czech Republic, Estonia, Hungary, Latvia, Lithuania, Poland, Slovakia and Slovenia*, Food and Agriculture Organization, Rome.

Fink, M., Grajewski, R., Siebel, R. and Zierold, K. (1994) Rural women in East Germany, in D. Symes and A.J. Jansen (eds.) *Agricultural Restructuring and Rural Change in Europe*, Wageningen University Press, Wageningen, pp.282-295.

Forsberg G., Gonäs, L. and Perrons, D. (2000) Paid work: Participation, inclusion and liberation, in S.S. Duncan and B. Pfau-Effinger (eds.) *Gender, Economy and Culture in the European Union*, Routledge, London, pp.27-48.

Fotev, G., Draganova, M., Nmedelcheva, T. and Molhov, M. (2001) *Bulgarian Rural Women Today*, Bulgarian Academy of Science Institute of Sociology, Sofia.

Franco, A. and Blöndal, L. (2003) Labour Force Survey, principal results 2002 acceding countries, *Statistics in Focus – Population and Social Conditions 16*, Eurostat, Office for Official Publications of the European Communities, Luxembourg.

Franco, A. and Jouhette, S. (2003) Labour Force Survey, principal results 2002 EU and EFTA countries, *Statistics in Focus – Population and Social Conditions 15*, Eurostat, Office for Official Publications of the European Communities, Luxembourg.

Franco, A. and Winqvist, K. (2002a) Women and men reconciling work and family life, *Statistics in Focus - Population and Social Conditions 9*, Eurostat, Office for Official Publications of the European Communities, Luxembourg.

Franco, A. and Winqvist, K. (2002b) The entrepreneurial gap between women and men,

Statistics in Focus - Population and Social Conditions 11, Eurostat, Office for Official Publications of the European Communities, Luxembourg.

Gidarakou, I. (1999) Young women's attitudes towards agriculture and women's new roles in the Greek countryside: A first approach, *Journal of Rural Studies*, 15, pp.147-158.

Giovarelli, R. and Duncan, J. (1999) *Women and land in Eastern Europe and Central Asia*, paper presented at the conference on 'Women farmers: Enhancing rights and productivity', 26–27 August 1999, Bonn.

Gourdomichalis, A. (1991) Women and the reproduction of family farms: Change and continuity in the region of Thessaly, Greece, *Journal of Rural Studies*, 7, pp.57-62.

Halliday, J. and Little, J. (2001) Amongst women: Exploring the reality of rural childcare, *Sociologia Ruralis*, 41, pp.423-437.

Haugen, M.S. (2004) Rural women's employment opportunities and constraints: The Norwegian case, in H. Buller and K. Hoggart (eds.) *Women in the European Countryside*, Ashgate, Aldershot, 59-82.

Haugen, M.S. and Villa, M. (2003) *The countryside as a rural idyll or a boring place? Young people's images of the rural*, paper presented at the XXth Congress of the European Society of Rural Sociology, 18-22 August 2003, Sligo.

Haukanes, H. (2001, ed.) *Women After Communism: Ideal Images and Real Lives*, University of Bergen Centre for Women's and Gender Research, Bergen.

Henderson S.R. and Hoggart, K. (2003) Ruralities and gender divisions of labour in Eastern England, *Sociologia Ruralis*, 43, pp.349-378.

Högbacka, R. (1999) *Women's work and life-modes in rural Finland: Change and continuity*, paper presented at the conference on 'Gender and Rural Transformations in Europe', 14-17 October 1999, Wageningen University.

Hoven, van B. (2004) Rural women in the former GDR: A generation lost?, in H. Buller and K. Hoggart (eds.) *Women in the European Countryside*, Ashgate, Aldershot, pp.123-140.

Irmen, E. (1997) Employment and population dynamics in OECD countries: An intraregional approach, in R.D. Bollman and J.M. Bryden (eds.) *Rural Employment: An International Perspective*, CAB International, Wallingford, pp.22-35.

Jervell, A.M. (1999) Changing patterns of family farming and pluriactivity, *Sociologia Ruralis*, 39, pp.100-115.

Kelly R. and Shortall, s. (2002) 'Farmers' wives': women who are off-farm breadwinners and the implications for on-farm gender relations, *Journal of Sociology*, 38, pp.327-343.

Kligman, G. (1994) The social legacy of communism: Women, children and the feminization of poverty, in J.R. Millar and S.L. Wolchik (eds.) *The Social Legacy of Communism*, Cambridge University Press, Cambridge, pp.252-270.

Knudsen, K. and Waerness, K. (2001) National context, individual characteristics and attitudes on mother's employment: A comparative analysis of Great Britain, Sweden and Norway, *Acta Sociologica*, 44, pp.67-79.

Kulcsar, L. and Verbole, A. (1997) *National Action Plans for the Integration of Rural Women in Development: Case Studies in Hungary and Slovenia*, Food and Agriculture Organization, Rome.

Laas, A. (1999) *New providers and carers – winners or losers*, paper presented at the conference on 'Gender and Rural Transformations in Europe', 14-17 October 1999, Wageningen University.

Lindsay, C., McCracken, M. and McQuaid, R.W. (2003) Unemployment duration and employability in remote rural labour markets, *Journal of Rural Studies*, 19, pp.187-200.

Little, J. (1991) Theoretical issues of women's non-agricultural employment in rural areas, with illustrations from the UK, *Journal of Rural Studies*, 7, pp.99-105.

Little, J. (1994) Gender relations and the rural labour process, in S.J. Whatmore, T.K.

Marsden and P.D. Lowe (eds.) *Gender and Rurality*, David Fulton, London, pp.11-29.

Little, J. (1997a) Employment marginality and women's self-identity, in P.J. Cloke and J. Little (eds.) *Contested Countryside Cultures*, Routledge, London, pp.138-157.

Little, J. (1997b) Constructions of rural women's voluntary work, *Gender, Place and Culture*, 4, pp.197-209.

Luxton, M. (1997) The UN, women and household labour: Measuring and valuing unpaid work, *Women's Studies International Forum*, 10, pp.431-439.

Majerová, V. (1999) *Development of labour market in the Czech countryside: Expectations of rural women*, paper presented at the conference on 'Gender and Rural Transformations in Europe', 14-17 October 1999, Wageningen University.

Mauthner, N., McKee, L. and Strell, M. (2001) *Work and Family Life in Rural Communities*, York Publishing Services, York.

Melis, A. (2003) Women participating in active labour market policies. *Statistics in Focus – Population and Social Conditions 17*, Eurostat, Office for Official Publications of the European Communities, Luxembourg.

Meurs, M. (1999) *Economic strategies of surviving post-socialism: Changing household economies and gender division of labour in the Bulgarian transition*, paper presented at the conference on 'Gender and Rural Transformations in Europe', 14-17 October 1999, Wageningen University.

Meyer, H. von (1997) Rural employment in OECD countries: Structure and dynamics of regional labour markets, in R.D. Bollman and J.M. Bryden (eds.) *Rural Employment: An International Perspective*, CAB International, Wallingford, pp.3-21.

Momsen, J., Kukorelli Szorenyi, I. and Timàr, J. (1999) *Regional differences in women's rural entrepreneurship in Hungary*, paper presented at the conference on 'Gender and Rural Transformations in Europe', 14-17 October 1999, Wageningen University.

Morell, I.A. (1999) *Post-socialist rural transformation and gender construction processes*, paper presented at the conference on 'Gender and Rural Transformations in Europe', 14-17 October 1999, Wageningen University.

Moss, J.E., Jack, C.G., Wallace, M. and McErlean, S.A. (2000) *Securing the future of small farm families: The off-farm solution*, paper presented at the conference on 'European Rural Policy at the Crossroads', 29 June – 1 July 2000, Arkleton Centre for Rural Development Research, University of Aberdeen.

Nì Laoire, C. (2001) A matter of life and death? Men, masculinities and staying 'behind' in rural Ireland, *Sociologia Ruralis*, 41, pp.220-236.

Oedl-Wieser, T. (1999) *Challenges for a gender-sensitive rural policy: Analysis of the impacts of EU-structural funds interventions for women in rural areas in Austria*, paper presented at the conference on 'Gender and Rural Transformations in Europe', 14-17 October 1999, Wageningen University.

Ogbor, J.O. (2000) Mythicizing and reification in entrepreneurial discourse: Ideology-critique of entrepreneurial studies, *Journal of Management Studies*, 37, pp.605-635.

O'Hara, P. (1998) *Partners in Production: Women, Farm and Family in Ireland*, Berghahn, Oxford.

Oldrup, H. (1999) Women working off the farm: Reconstructing gender identity in Danish agriculture, *Sociologia Ruralis*, 39, pp.343-358.

Olsson, H. (2000) *Social Security, Gender Equality and Economic Growth*, Swedish Presidency report (http://europea.eu.int./comm/employment_social/equ_opp/information_eu.html#temp).

Overbeek, G. (1999) *Labour situation of farm women in Europe and the Netherlands*, paper presented at the conference on 'Gender and Rural Transformations in Europe', 14-17 October 1999, Wageningen University.

Overbeek, G., Efstratoglou, S., Haugen, M.S. and Saraceno, E. (1998) *Labour Situation and*

Strategies of Farm Women in Diversified Rural Areas of Europe, Office for Official Publications of the European Communities, Luxembourg.

Perrons, D. and Gonäs, L. (1998) Perspective on gender inequality in European employment, *European Urban and Regional Studies*, 5, pp.5-12.

Persson, L.O. and Westholm, E. (1993) Turmoil in welfare system reshapes rural Sweden, *Journal of Rural Studies*, 9, pp.397-404.

Pfau-Effinger, B. (1994) The gender contract and part-time work by women: Finland and Germany compared, *Environment and Planning*, A26, pp.1355-1376.

Pfau-Effinger, B. (2000) Conclusion: Gender cultures, gender arrangements and social change in the European context, in S.S. Duncan and B. Pfau-Effinger (eds.) *Gender, Economy and Culture in the European Union*, Routledge, London, pp.262-276.

Pine, F. (2001) 'Who better than your mother?' Some problems with gender issues in rural Poland, in H. Haukanes (ed.) *Women After Communism: Ideal Images and Real Lives*, University of Bergen Centre for Women's and Gender Research, Bergen, pp.51-66.

Pine, F. (2002) Retreat to the household? Gendered domains in postsocialist Poland, in C.M. Hann (ed.) *Postsocialism: Ideals, Ideologies and Practices in Eurasia*, Routledge, London, pp.95-113.

Plantenga, J. and Rubery, J. (1999) Introduction and summary of main results, in J. Plantenga and J. Rubery (eds.) *Women and Work: Report on Existing Research in the European Union*, Office for Official Publications of the European Communities, Luxembourg, pp.1-12.

Post, J. and Terluin, I. (1997) The changing role of agriculture in rural employment, in R.D. Bollman and J.M. Bryden (eds.) *Rural Employment: An International Perspective*, CAB International, Wallingford, pp.305-326.

Rangelova, R. (1999) *Rural restructuring and gender dimensions of socio-economic change in Bulgaria*, paper presented at the conference 'Gender and Rural Transformations in Europe', 14-17 October 1999, Wageningen University.

Rooij, S. de and Bock, B.B. (forthcoming) Rural women and food security in Europe: Facts, figures and trends, in: FAO (eds.) *Rural Women and Food Security: Current Situation and Perspectives* [working title], Food and Agriculture Organization of the United Nations, Rome.

Rubery, J. (1999) Overview and comparative studies, in J. Plantenga and J. Rubery (eds.) *Women and Work: Report on Existing Research in the European Union*, Office for Official Publications of the European Communities, Luxembourg, pp.13-26.

Rueschemeyer, M. (1998, ed.) *Women in the Politics of Postcommunist Eastern Europe*, revised and expanded edition, M.E. Sharpe, Armonk, New York.

Rural Development Committee (2001) *Report on the Impact of Changing Employment Patterns in Rural Scotland - Volume Two*, Scottish Parliament Paper 254, Edinburgh (http://www.scottish.parliament.uk/offical_report/cttee/rural-01/rar01-01-v02-01.htm).

Sackmann, R. and Häussermann, H. (1994) Do regions matter? Regional differences in female labour-market participation in Germany, *Environment and Planning*, A26, pp.1377-1396.

Safiliou, C. (2001) *Men and Women Smallholders Across Europe*, synthetical international report for SU DG-XII on social exclusion of women smallholders across Europe, National Centre for Social Research (EKKE), Athens.

Terluin, I.J. and Post, J.H. (2000, eds.) *Employment Dynamics in Rural Europe*, CAB International, Wallingford.

Tijdens, K. (2002) Gender roles and labour use strategies: Women's part-time work in the European Union, *Feminist Economics*, 8(1), pp.71-99.

Trifiletti, R. (1999) Women's labour market participation and the reconciliation of work and family life in Italy, in L. den Dulk, A. van Doorne-Huiskes and J. Schippers (eds.)

Work-Family Arrangements in Europe, Thela Thesis, Amsterdam, pp.75-102.

UNDP (1999) *Transition 1999: Regional Human Development Report for Central and Eastern Europe and the CIS*, United Nations Development Programme Regional Bureau for Europe and the CIS, New York.

Verbole, A. (1999) *Negotiating Rural Tourism Development at the Local Level: A Case-Study in Pisece, Slovenia*, PhD thesis, Wageningen University.

Warren-Smith, I., Monk, A. and Parsons, S. (2001) *Women in micro-businesses: Pin money or economic sustainability for rural areas?*, paper presented at the conference on 'The New Challenge of Women's Role in Rural Areas', 4–6 October 2001, Agricultural Research Institute, Nicosia, Cyprus.

Weise, C., Bachtler, J., Downes, R., McMaster, I. and Toepel, K. (2001) *The Impact of EU Enlargement on Cohesion*, DIW (German Institute for Economic Research) and EPRC (European Policies Research Centre), Berlin and Glasgow.

Chapter 3

Women in Rural France: Mediators or Agents of Change?[1]

Marion Demossier

Introduction

While there has been a great deal of research on women farmers in rural France,[2] very few studies have been devoted to the position of women[3] living in rural areas who are not directly connected to the agricultural economy. The absence of such research can be partly explained by the dominant position of agricultural activities in French rural space and, by the same token, in French debates, but also by the prevailing influence of male farmers in the social and political fabric, which together have obscured the presence of women. It is also true that, despite a proliferation of research on rural France over the last 30 years, key contemporary issues facing women in rural areas have been completely neglected. However, recent international and European debates around the parity and position of women in relation to employment have put the issue of equality on the agenda. In this context, questions have arisen in relation to the rights and conditions of women living in France's rural areas.

Several recent initiatives headed by the European Commission and the Directorate-General for Agriculture have emphasized the active role women might play in European rural development. As part of the European Commission's policy of promoting equal opportunities, which was originally provided for in Article 119 of the Treaty of Rome, and which establishes the principle of equal pay for men and women, women's vital contribution to rural development is presented as an important priority 'to maximize human resources in maintaining the social fabric of rural communities and revitalizing local communities' (European Commission Directorate General for Agriculture, 2000, p.4). Thus, the position of women in

[1]   I should like to thank INSEE (Dijon) and Service des Droits des Femmes (Paris) for their help in preparing this chapter. I should also like to thank my French colleague, Anne Thierry, for providing some materials for this chapter.
[2]   For a survey of research on farm women in France, see Demossier (2000).
[3]   For an introduction to this topic, see Toutain (1998).

rural society has experienced a radical shift in terms of role, image and expectations. Women seem to be increasingly perceived as a crucial social category in the transformations affecting the countryside.

This chapter argues that not only have women already been part of the changes affecting French rural society, but they have contributed to the development of a culturally more urbanized and economically liberalized French society. Their role has been that of mediators living in the village, but working outside and adapting to modernity, thus operating as intermediaries between several worlds, be it agricultural and non-agricultural, urban or rural. For some, this is not without a certain cost, as they risk being seen by more traditional social groups in villages (retired farmers, local hunters, rooted families) as a potential source of tension and disorder, representing a modernity they find difficult to accept. This chapter aims to assess the complex nature of their position in social, cultural and economic terms, and to analyze to what extent their contribution remains original and integral to the numerous changes affecting French rural society.

In 1984, Henri Mendras wrote that rural France had been affected by major transformations which could be summarized as follows: the noisy decline and death of the peasantry, the rebirth of rural society, new uses of the countryside, which have become matters of lifestyle as much as agricultural production, the urbanization of rural society, and, finally, the disappearance of the traditional rural family. It is undeniable that the 1980s saw large-scale restructuring, not least because of the effects of reform of the Common Agricultural Policy. In just 20 years, the agriculturally active population in France decreased from four million to less than one million. Now only five per cent of the active population is employed in agriculture, with the French Census of 1999 classing only 642 167 individuals as farmers. Over the last 40 years, the number of farms has drastically reduced while their size has nearly tripled (14 hectares in 1955 against 40 today) and most farmers have become managers (Dubois Fresney, 2002, p.44). The rural exodus has left an ageing and declining number of agricultural males in the village, while young people, and especially young women (between 20 and 29 years old), have left the countryside for an urban life. Another consequence is the increase in the number of unmarried farmers, which is seen as a serious problem for the social life of both the village and the farm. This also had the indirect result of encouraging more farmers to marry women from outside the agricultural community, who have brought a supplementary salary to the household, and thereby transformed the nature of the rural family. But despite the decline of the agricultural industry, with 90 per cent of households living in 'dominantly rural areas' (see below), which are not composed of any agricultural workers, and with less than 20 per cent of rural jobs classed as agricultural (INSEE-INRA, 1998, p.6), agriculture still plays a major political and social role in village life. Agriculture in this sense is defined not only to mean farming activities but also the ownership of land, family lineages rooted in the village and the various categories of actors who have a knowledge of the countryside (retired farmers, hunters, etc.).

Following the major changes affecting rural France, the position of women in rural areas has also witnessed a radical transformation. First, women's participation in French agriculture has declined, with women increasingly working outside the family farm (Braithwaite, 1994, p.59). In addition, more women in rural areas have become visible and economically active. They are often described as a 'vital human resource'.[4] But is this an accurate description of the complex and evolving experiences of rural women?

A full review of their situation remains a difficult task, as national statistics have failed to integrate fully the gender factor into their demographic, economic and social compilations. The state of research, in general, remains fragmented and heterogeneous. It is therefore difficult to draw firm conclusions about the situation of women in rural France. As a consequence, what this chapter aims to provide is insight into the experiences of women in rural life, as well as to highlight areas that are in need of further research. The French example could also contribute to the wider debate about the changing position of women in rural areas in Europe as a whole.

Rural France: Towards an Urbanized Society?

The evolution of the statistical definition of 'rural France' illustrates the need to treat any generalizations about the role of women in rural areas with caution. In the first instance, it is necessary to define what we mean by 'rural France', before turning our attention to women living in rural areas. On account of its history and culture, France is one of the few European countries to have clearly defined the concept of the rural zone. The INSEE (Institut National des Statistiques et Etudes Economiques) until recently classed any community of less than 2 000 inhabitants as rural. These rural zones included more than 31 000 communes out of the national total of 36 000 and represented approximately 85 per cent of French territory. This definition excluded the *bourgs* that were the dynamic part of rural space, but it did include communes under the strong influence of nearby towns. Despite their imperfect nature, these statistics gave a broad indication of the various territories comprising rural France. However, they did not take into account the recent movement of population between different parts of French rural and urban space. Most commentators qualify this new demographic trend as *rurbanization*.

Major changes have indeed affected the relationship between rural and urban parts of France to the extent that new statistical tools have been created by the State to measure their development. Working in a nearby town and living in the countryside is a major and relatively recent phenomenon, which affects around two million workers who live in peri-urban communes and a further 750 000 living in 'dominantly rural areas'. Today, these 'dominantly rural areas' (a category created from the census of 1990), which is a new category whose title captures the difficulty of separating rural from urban, account for 70 per cent of France and include 25 per cent of the overall population (Dubois Fresney, 2002, p.12). In 1999, it accounted for 13.6 million inhabitants. If this new category is applied retrospectively (including the same communes), it reveals that the population is

---

[4]   http://europa.eu.int/comm/archives/leader2/rural-en/biblio/women/art03.htm.

almost the same as in 1962 (Bessy-Pietri *et al.*, 2000). Yet these areas gained 247 000 inhabitants in the nine years from 1990 to 1999. According to Bessy-Pietri, Hilal and Schmitt (2000), this demographic upturn, which is due to a higher migratory influx than any natural deficit, has spread out to reach most communes. The migratory balance has become positive, even in isolated rural areas. In addition to which, there has been a sharp rise in rural inhabitants living near growing urban areas, especially on France's western and southern borders, in the greater western suburbs of Paris and in the regions of Alsace, Midi-Pyrénées and Rhône-Alpes. When the urban structure is weak or in decline, rural population growth is generally negative (Bessy-Pietri *et al.*, 2000, p.4). This phenomenon has revealed a parallel evolution in demographic and economic terms between urban and rural areas, which poses a real threat to the future planning of the French territory as a whole, and which affects the traditional nature of rural France.

Of the changes confronting French rural society, the issue of *ruralité* has recently emerged as a new object of intense contestation, due partly to its evolving nature and to the changing relationship between 'countryside' and 'town' (see Clout and Demossier, 2003). The boundaries between these two worlds have blurred, and consequently, 'agricultural' is no longer synonymous with 'rural'. As many scholars have noted, far bigger issues are at stake for rural France. Long-distance commuting, widespread retirement to the countryside, the proliferation of *résidences secondaires* and the growing popularity of rural tourism are amongst the cherished commodities increasingly consumed by urban dwellers. According to the historian Jean-Pierre Rioux (2003, p.931), contemporary rural France is associated with a new form of *ruralism*, which is headed by the urban part of the population. But the integration of these two traditionally separate worlds has not yet been achieved. Significantly here, the position of women is crucial in the major cultural transformation through which the rural territory becomes urbanized.

Since the economic crisis of the 1970s, the agricultural world has undergone a major transition with the development of multiple activities and agro-tourism. In many cases, farmers' wives, who a generation ago were involved in *petit commerce*, or who were helping out on the farm, are now working as employees elsewhere.[5] These women have actively participated in the transformation affecting rural space and, increasingly, they have been responsible for breaking down frontiers separating rural France from its urban counterpart. Various activities, from bed and breakfast[6] to the revival of local cheese production, have been part of this wider movement affecting rural France. Tourism, education and other rural activities have transformed the rural space and its inhabitants into a commodity, and the role of women in this change has to be fully recognized, as they were traditionally more open to the outside world than men.

---

[5] INSEE (1993) quoting the agricultural census of 1970 and 1988, Table 2, p.27. There has not been an update of these statistics.

[6] Between 1980 and 1988, the number of guest houses and holiday homes rose fourfold while the number of rooms and gîtes doubled. For more details, see INSEE-INRA (1998, p.86).

Another feature of this trend towards greater contact with the outside world is a development of activities alongside the farm that encourage links with town people, as is the case with agro-tourism or viticulture. The sale to the public of farm products and the growth of rural tourism are particularly pertinent. In the majority of cases, women have been responsible for the development of these sectors and they exercise direct control over the organization involved (GREP, 1998). Again, disparities in the nature of these activities impact on the definition of women's position. For instance, in viticulture, despite the fact that women are involved in the commercial side of the business, in general they are still not invited to take important visitors for wine-tasting in the cellar,[7] therefore their role remains secondary. Even so, as a great majority of women in rural areas have started to join the labour force, especially in the tertiary sector, they have enjoyed more financial freedom and have become more urbanized in their values and lifestyles. That said, living in rural France today remains a distinctive experience.

Women in Rural and Urban France: Similarities and Differences

The situation of women, notably of women farmers, in rural France remains on many levels separate from that of women living in towns and cities. Nevertheless, this divide has been reduced. According to Xavier Toutain (1998, p.18), '... concerning professional activity, the situation of women in rural France has been drastically transformed and has become similar to that of women in urban France'. In 1997, the activity rate of young women in the rural areas of France reached 78 per cent for the first time, so matching that of the same age group of women in urban areas. The proportion of economically active women employed in the tertiary sector in rural areas of France was 72 per cent in 1997, up from 62 per cent in 1990 (European Commission Directorate General for Agriculture, 2000, p.7). This tendency is confirmed by the recent 2002 activity rate, which distinguishes the various categories of communes. The rate of activity for women living in a rural commune (aged 25 to 49 years) is 81.9 per cent, while it is 82.2 per cent for Paris and its suburbs and 79.8 per cent for the population overall. These statistics remain lower than the same age range for men (25-49 years), with respectively 96.3 per cent in rural communes, 95.0 per cent in Paris and its suburbs and 94.7 per cent overall. It could be concluded that more women are now joining the labour force and differences are vanishing. However, their position in relation to men is still unequal (see Gregory, 2000; Windebank, 2000).

Moreover, national statistics have shown that women are generally working part-time, with 83 per cent of part-time jobs occupied by women. Since 1982, a series of legislative measures have encouraged these forms of activity. It is also true that specific economic sectors, such as small businesses and shops, restaurants, domestic service, canteens and tourism, are associated with flexible hours and part-

---

[7] There are more women involved in viticulture and the wine industry than used to be the case. However, some traditional taboos exist, which prevent them from being fully recognized.

time work (INSEE, 1995, p.132). These sectors are very often found in rural areas. Similarly, seasonal and informal activities are not fully taken into account when assessing their contribution to the agricultural and tourist sectors. It is also important to note that the position of working women is considered more flexible, but is also more precarious and unequal, especially in uncertain economic contexts (35 hours, variability of working hours, etc...). This applies to rural as well as urban women. It could be concluded that there is still progress to be made before French women achieve real equality in professional terms.

For many commentators, the general situation of women in rural areas has improved and their lifestyle, nowadays, is very similar to that of women living in cities and towns. The most frequently cited example of this is leisure and cultural activities. The new willingness to 'go out', whether to restaurants, the cinema or the theatre, as well as the reduction in men's time spent in cafes, as opposed to watching television, illustrates the convergence of agricultural and urban lifestyles. Even farmers have embraced the once taboo concept of a vacation (Demossier, 2000, p.199). In 1994, official figures revealed that the number of holidays taken by families living in rural zones had tripled compared to the situation in 1982. It is, however, important to note that there are still some sharp differences between rural and urban dwellers in terms of holiday destinations: Parisians go abroad more often than people living in rural communes (21 per cent against 14 per cent) and they stay away longer (with an average of 17 nights as against just 11 nights for rural residents). In winter, the majority of inhabitants in rural communes prefer to go to towns and cities (Roquette, 2003), rather than the ski slopes.

For economically active rural women, it remains more difficult than for their urban counterparts to reconcile professional and domestic occupations. For farm women, the amount of time that needs to be devoted to the household and other domestic tasks is greater because of their connection to work on the farm (Demossier, 2000, p.203). The 'Enquête emploi du temps' (survey of daily timetables of French people) for 1998-1999 gives a detailed account of the daily activities undertaken by men and women in relation to the areas in which they live. In both rural communes and in communes of less than 20 000 inhabitants, women (economically active or not) devote more time than men to the education of their children and to the maintenance of their household. In terms of time devoted to professional activities and training, men devote a greater amount of time to work and training than women, be they economically active or not. Another indicator of the difficult position occupied by women in rural France is the free time that each category enjoys. Men have on average 223 minutes of free time per day while women have 184 and this difference is more acute when they are both classified as economically inactive (64 points of difference compared to 39 when both are economically active). Results for other demographic categories demonstrate that women in rural areas devote more time to education and domestic activities than their urban counterparts.

Thus, the general convergence of lifestyles seems to affect only one part of the population of rural France, while the other part could be characterized as less urbanized and more traditional. Depending on the nature of rural communes, especially on whether they are sufficiently or insufficiently equipped (in terms of infrastructure), on the type of social fabric (which group dominates village life) and on the socio-economic position of individuals who have made the choice of living in rural areas, the picture can be drastically different. If we were to compare the isolated Ardèches with the urbanized and wealthy villages of Burgundian wine-growers, then very different experiences of being a woman in rural France would have to be addressed. The position of women in rural France also depends upon their numbers and social position, and, thereby, on their civic participation at a local level. Again, regional diversity, the nature of the rural fabric, the situation of the individual (married, divorced or single) and many other factors affect their participation in social and economic life.

Although women outnumber men in the population as a whole, this is not the case for all ages and for all categories of territories (INSEE, 2001, p.10). For one, the proportion of women in 'dominantly urban space' increases as urbanization develops, while it diminishes in 'dominantly rural space' and especially in the category 'isolated rural' (INSEE, 1998, p.41). It is equally important to point out that, among families who have recently moved to the rural part of France, a distinct profile has emerged, as this group has a profile with many young couples with children. Within these migration inflows, it is very often qualified workers and *professions intermédiaires*[8] that are over-represented. 'Isolated rural areas', on the other hand, are characterized by inflows that are marked by the arrival of older couples, couples without children and single people (INSEE-INRA, 1998, p.46). Yet what these inflows have helped achieve is a changed balance in the economic base of the rural population, for workers and employees are now three times more numerous than agricultural workers and farmers in rural areas (Sylvestre, 2002, p.8). This 'renaissance rurale' has radically altered the traditional world of the village (Kayser, 1990).

These factors need to be borne in mind when we examine the issue of unemployment, as the situation could drastically differ from one area to another, as well as from one generation to the next. In 1996, the standard of living of rural communes was some 20 per cent below the national average (Lagarenne and Tabard, 1998, p.1). In the city, as in the countryside, around 40 per cent of RMIstes (a 'safety net' financial benefit designed to help the most deprived) are single men, 20 per cent single women, 20 per cent single women with children and the last 20 per cent is composed of couples with or without children (INSEE-INRA, 1998, p.72). If we examine the rate of unemployment in detail, in the 15-24 age group, 34 per cent of women and 19 per cent of men were unemployed in the rural areas of France in 1997, which is a much greater gap than amongst the same age group in urban areas (European Commission Directorate General for Agriculture, 2000, p.7). In 2002, the percentage is 32.0% for women and 24.9% for men (see Table 3.1). For the category 15-64 years of age, the female unemployment rate is the same for rural communes as it is for Paris and its suburbs, while remaining lower than the national average.

---

[8]   The designation 'professions intermédiaires' includes middle-ranking employees in administration, primary school teachers, technicians and supervisors.

Table 3.1 Unemployment rate by age and category of French commune

Commune type - inhabitants	Worker age (years)						
	15-19	20-24	25-49	50-54	55-59	60-64	15-64
Rural Communes	22.4	19.6	8.0	6.3	4.9	1.9	8.2
UU (Urban Units) of < 20 000	31.7	25.0	10.8	9.5	5.9	5.5	11.7
UU of 20 000 to 200 000	39.5	27.9	11.5	9.4	6.4	1.5	12.4
UU of more than 200 000	30.6	20.5	11.0	8.9	5.5	7.3	11.2
Paris and suburbs	26.1	15.0	7.6	8.3	7.4	4.4	8.2
Total	30.0	21.9	9.6	8.3	5.9	3.8	10.1

Source: INSEE, Enquête-Emploi 2002.

It is also worth mentioning that men are less likely to be unemployed than women in rural communes (5.2 per cent against 8.2 per cent across all ages). Thus, there are differences between age groups, with younger rural women more affected by unemployment and by the nature of the rural area they live in. In 1994, women formed more than half of those from the rural population seeking a job (Braithwaite, 1994, p.77). This general situation still prevails, despite a slight improvement. Both the European Commission and local associations have launched initiatives that have tried to address the complex issue of work and training in the context of the policies of decentralization pursued by the French State, which are aiming to shift financial responsibilities to regions and communes. The review of these projects still remains incomplete. As we can see, even if there is a convergence in some specific areas between rural and urban women, differences remain which make the experience of rural women in France quite distinctive.

Table 3.2 Men and women by category of French commune (in thousands)

Commune type – inhabitants	1990		1999	
	Women	Men	Women	Men
Rural Communes	6 856	6 892	7 146	7 179
UU (Urban Units) of < 20 000	4 860	4 631	5 088	4 820
UU of 20 000 to 200 000	4 946	4 586	5 060	4 651
UU of more than 200 000	7 503	6 881	7 819	7 115
Paris and suburbs	4 892	4 578	4 990	4 654
Total	29 057	27 568	30 101	28 419

Source: INSEE (2001, p.10).

Non-Agricultural and Agricultural Women in Rural France

The last Census of 1999 indicated that 7 146 000 women and 7 179 000 men lived in rural communes (Table 3.2), showing a slight increase since 1990. Among them, 971 000 individuals were occupied in the agricultural sector in 2000, compared to 1 394 000 in 1995.[9] When using the term 'rural women', it is necessary to acknowledge the striking diversity of situations encountered in terms of social and economic status, as well as professional and family cultures. As underlined by the European Commission, 'rural women too are not a homogeneous group' (European Commission Directorate General for Agriculture, 2000, p.4) and this is more striking in the case of France, which is characterized by strong regional diversity. Women have different roles and occupations, on farms and in family businesses, in employment and in community activities. Their needs and interests differ too, particularly from one group to another, and depending on the size and composition of their family and the age(s) of their children. Again, the economic and social changes that they have to face do not affect them in the same way, as they offer opportunities for some while bringing difficult challenges for others.

According to Franz Fischler (1996, p.4), European Commissioner with responsibility for Agriculture and Rural Development, 'women living in rural areas encounter several disadvantages: their employment opportunities are limited, child-minding facilities are inadequate, communal transport in rural areas rarely meets demand, training centres are scarce'. While the majority of the rural population are owner occupiers and pay less for accommodation, they have to spend more time and money for travel and to communicate than their urban counterparts (INSEE-INRA, 1998, p.8). Living in a rural commune means that the level of amenities is less satisfactory than in a town. Many primary schools do not have computers and many small villages do not have transport facilities, and often lack basic services, such as post offices or village shops. Again, major differences prevail between one commune and another. Because of all these difficult conditions and their mediating position between rural and urban France, women, and especially the *neo-ruraux*, have been very influential in implementing the changes needed to accomplish the modernization of their regions.

Since the 1990s, initiatives sponsored by the European Union have provided new opportunities that have been seized upon by some French regions, such as Tarn or Gard, to facilitate women's access to employment. It is clear that recent European programmes, such as *Now* or *Leader 1*, have launched several initiatives with regional associations and, consequently, have helped create a more favourable environment for some women. However, some of these opportunities have to be relayed by the French economic fabric. The dynamism of specific rural areas has, to some extent, created the necessary conditions for diverse initiatives aimed at promoting women's participation in the labour force. Some villages have tried to encourage their integration and have helped women to find jobs in childcare or in

---

[9]   For the same periods, there were 670 000 men in 2000 and 914 000 in 1990 (Eurostat, 2002, pp.98-99) for 301 000 women in 2000 and 480 000 in 1990.

local associations. A similar idea lay behind a training programme in Haute-Vienne (Braithwaite, 1996, p.3), which was aimed at enabling rural women to become bus drivers to highlight the mobility problems of those without personal transport. However, their participation often remains voluntary, is not always officially declared or is on a temporary basis.

In the publication *Village Magazine* (Number 3, February 2000), the issue of equality between men and women in rural France was addressed by presenting 115 women and their experiences of living in rural areas. These women are not portrayed as exceptional cases, merely as examples of innovation and dynamism that could be an inspiration for other women. These 115 stories were structured around three topics, reflecting aspirations of living in rural France: the love for their job, the idea of changing everything, and, finally, the need to learn for yourself and assist others. For most of these women, it was clear that the decision to contribute to village life or the environment in which they were living played a major role. Their experiences suggest that they have been successful in combining their quest for an individual identity with the vicissitudes of modern life. Involved in businesses, politics, tourism or the local community, women's contributions to the new dynamism of the French countryside was presented in this publication as both essential and outward looking.

In a similar vein, the publication *Femmes en milieu rural: Des initiatives pour l'animation et le développement des services et l'emploi* has anticipated the wide range of activities now covered by women in rural France (SEGESA-Service des Droits des Femmes, 1995). The initiatives presented in this brochure combined local development, community needs and female employment or training. In Aquitaine, for example, in 1993-1994, 15 women followed a training course designed to provide them with specific competencies and a project concerning the creation of new services and activities, such as the opening of a bar-restaurant in a small neighbouring village. These initiatives also involved a reinforcement of female competencies in the maritime sector of the Finistère. Without a recognized professional status, most of the fishermen's spouses demanded greater visibility, recognition for their contribution to the commercial side of the business and specific training. Grounded within these demands, a project emerged amongst themselves to fulfil their need for complementing their husbands and ensuring recognition of their participation. Their project was backed by a local association and has been supported by aid from European programmes.

In an article published in 1998, Dominique Massé (1998, p.100) recalls the struggle of the Group 'Terre et Mer' to obtain social recognition of the status of spouse in the small fishing and shellfish farming industry. The group was obliged to spend a year negotiating with national representatives to obtain the law of 19 November 1997 (on the same basis obtained by farmer's wives in 1980), which gave them full recognition. The initial project, started in Brittany and in Arcachon, was able to generate discussion amongst the women about the professional status they wanted to have in relation to their husbands and the farm: 'The differences between women, in relation to their activities, the place of this activity, their daily lives were identified for the first time because their meeting has taken place around one consensual claim associated with their social status. The discussions have been intense, but they have been helpful in defining common and clear objectives' (Massé, 1998, p.100). The major outcome of this project was not only the recognition of their status, but the capacity of these women to engage in the legislative debate through various committees and to impose their views on a previously unidentified issue. By their capacity to negotiate at different levels, they have been transformed into agents of change.

Among all the examples cited, it is clear that more women want to become independent and run a business. In France, 60 per cent of family businesses – mainly craft and commercial enterprises – are run by a couple. Only 6 per cent of female spouses have the status of 'joint collaborator'. A similarly small percentage is classed as partners or are employed as associates. The great majority often work more than 39 hours per week, particularly in small commercial businesses, without any legal status or social protection. This is a true hidden economy, which does not help the recognition of women's situation (European Commission Directorate General for Agriculture, 2000, p.8). Recent research conducted in Rhône-Alpes has shown that, despite the major involvement of women, significant obstacles, especially of an administrative nature, remain, which are described as a veritable '*parcours du combattant*' (military manoeuvres referring to a masculine culture),[10] especially for women attempting to set up a business. Yet these various European projects, whatever their outcomes, have at least made women aware of their own economic importance. By the same token, they have enabled the creation of new social ties and networks between various categories of rural actors with common problems.

What has helped in this regard is shared connections between women. Rural communities are very often characterized by strong social ties between their members, and especially female members, which provides the basis for social solidarity. Over the last 30 years, the number of women participating in local associations has tripled. In rural areas, this participation represents an opportunity to become independent and to integrate fully into civic life. In the various projects initiated by European programmes, such as *Leader*, there has been some attempt to facilitate relationships between agricultural and non-agricultural women. The Lot-et-Garonne Group (Aquitaine), for example, has worked in partnership with the *Plurielles* association, which was set up by women and actively supports women farmers who are looking for new sources of income in connection with family-run farms, or who would like to take up employment outside the farm (Jouffe, 1996, p.1). Cases have multiplied over the years with more collaboration between different types of social actors. New types of social networks have thus emerged with specific competencies, which can also be illustrated by the growing role of women as members of associations.

Whether they belong to local associations or in the case of older women to a club du troisième âge (club for retired people), their participation is nearly always connected to social matters (Demossier, 2000, p.199). Women have traditionally been the initiators of local projects related to childcare (local crèches), canteens or

---

[10] See Confédération Paysanne (2001) *Campagnes solidaires*, n°152, mars. http://confederationpaysanne.fr/cs/152pointvue.htm.

school mergers in small communes. For example, in three communes of the Côte d'Or (Burgundy), the parents' association worked jointly with three mayors to revive local schools (*maternelle* and primary school) and attract more people to move into the area. Most participants in the project were women (working and non-working) involved in their children's education. At local level, women are more active in these associations and they are more numerous than men in parental and religious groups (Table 3.3). Thus, they play an active role in the areas traditionally related to their position as women and mothers.

Table 3.3 Percentage population participation in main types of association

	Women	Men
All associations	38.29	49.68
Associations or sports clubs	13.63	23.35
Cultural or musical associations	8.02	7.41
Clubs for retired people	5.30	4.03
Parents' association	4.34	2.02
Unions or professional groups	3.93	7.85
Religious or parish groups	3.78	2.25
Local associations	2.74	2.09

Source: Adapted from INSEE (2001, p.25).

Women in Rural France: Mediators or Agents of Change?

Because of their substantial involvement in the community and their multiple roles, rural women are generally less involved in decision-making at local or regional levels. Women farmers have rarely, if ever, participated in national or international politics and they have generally been excluded from the professional world of their husbands. Between 1979 and 1983, there were no women elected to the *Chambre d'Agriculture* in no less than 39 *départements* (Jacques-Jouvenot, 1997, p.106). As Braithwaite (1996, p.3) has noted, often domestic and community maintenance roles not only restrict women from participating in decision-making, but also means that they provide the support that enables men to participate. There are several examples that could be given of mayors of small communes who rely on their wives to assume various social and communal responsibilities. The unequal share of the domestic burden is a significant constraint on the participation of women, but this is not the only explanation, as rural France is still defined by a patriarchal and traditional type of social organization, whereby women, and especially working women, are perceived as secondary and apolitical. By entering into the political arena, women risk being accused of fomenting social disorder and of undermining the position of their husband in the eyes of the community. Their traditional roles within the family and in terms of supervising their children's education is by the same token subject to close and often critical scrutiny by the older generation.

In a recent report published by the CEPFAR, Herlitzka (1995) underlines the weak participation of women in agricultural and rural affairs. Women are underrepresented in both the National Assembly and the Senate, with respectively 1.4 per cent of women in the production and exchanges commission of the National Assembly and 2.5 per cent sitting on the economic affairs commission of the Senate. In the Ministry for Agricultural Affairs, their number declines as their status improves, with 67 per cent in category D while 36 per cent compose the category A.[11] This survey, which was conducted with the professional agricultural organizations in Europe, enables us to refine the overall picture of women's participation in political processes. The *Fédération Nationale des Syndicats d'Exploitants Agricoles* (FNSEA) has 240 000 women members (40 per cent of the total). Yet for *Unions Professionnelles Agricoles* (UPA) and *Union des Exploitants Familiaux* (UEF) only 3.6 per cent of women are in decision-making positions. Both sets of statistics illustrate the low involvement of women in decision-making. When they occupy a key position, further investigation reveals that in terms of the functions fulfilled, their role is essentially consultative and co-decisional. However, some changes do seem to be occurring at a professional and a local level.[12]

Several explanations could be advanced to explain women's lack of visibility in the sphere of French agriculture. Despite the fact that, in 1980, farmers' wives could obtain status in their own right as co-owner or manager of a farm, most women have a tendency to occupy a back seat role. The argument advanced is that a public function and family duties cannot coexist. Behind these justifications, the patriarchal organization of the agricultural world and the complementarities of men and women to represent the farm and defend the profession, remain the major obstacle to real equality (Herlitzka, 1995, p.44).

However, these obstacles are not only confined to the professional world, but are also a feature of French political life (for a full discussion, see Cross, 2000). France has been slow to develop in two key areas affecting women's rights; namely, in recognition of the right to vote (1944) and in the representation of women in elected assemblies. It is interesting to note that, in 2001, only 5.9 per cent of women were elected to the French Senate and 21.8 per cent to the *conseils municipaux*, while 40.2 per cent were elected to the European Parliament (Reynié, 2003, pp.449-451). In 1994, women provided just 17 per cent of those elected to the *conseils municipaux* amongst communes of less than 3 500 inhabitants. In 1996, they comprised around 22 per cent. Recent elections have seen a greater participation of women in local politics, with approximately 47.3 per cent elected in communes of between 3 500 and 30 000 inhabitants. These results are direct and

---

[11] Categories A and D refer to the status of civil servants, with senior civil servants belonging to category A.

[12] For instance, the *Groupe des Jeunes Professionnel de la Vigne* in Beaune (the association of Burgundian young wine-growers) has recently elected a young woman to represent it.

indirect consequences of two recent constitutional laws on parity in political life (N°99-596 of 9 July 1999 and N°2000-493 of 6 June 1999), which favour the equal access of women to the electoral list for communes of more than 3 500 inhabitants. If we look at the results of the *élections municipales* of March 2000 in more detail, it is clear that these laws had an impact even on communes with less than 3 500 inhabitants. In the 2 624 communes of more than 3 500 inhabitants, 38 106 women were elected *conseillers municipaux* (19 432 more than in 1995). Of the 34 150 communes of less than 3 500 inhabitants that did not have the same obligation, 118 321 women were elected to the *conseil municipal* (an increase of 29 594 compared to 1995; Du Granrut, 2002, p.10). However, these results need to be treated with caution as electoral results for mayors are less positive in terms of the increased participation of women, with 11.2 per cent for communes of less than 3 500 inhabitants and less than 7 per cent for the other category, despite an increase of 40 per cent compared to the last elections. For the *adjoint au maire* (deputy mayor) the percentage is around 15 per cent, which shows a progression of 25 to 30 per cent. Despite a sense of progress, there still remains a lot to be achieved.

Conclusion

Since the mid-1950s, the *révolution silencieuse*, an expression coined by Michel Debatisse (1963) to describe the modernization of French agriculture, has taken place accompanied by a major transformation of rural life. In this major rupture, the position of women has been largely overlooked or even denied. More than any other part of France, rural areas were for centuries the world of men, whether these be farmers, hunters, woodcutters, technicians, village *maires* or, more recently, ecologists. However, the 1990s have seen this picture changing, with recent publicity for the new countryside portraying a smiling woman welcoming tourists at the door of her house. During the United Nation's worldwide day devoted to women in rural areas (15 October 2002), 60 women farmers went to the European Parliament in Brussels to speak. The internationalization of the issue of women, following the United Nations fourth conference in 1997, has illustrated the extent to which France has started to become more integrated and liberalized. Several initiatives have started following this international conference and the issues of gender and parity have been debated. It is undeniable that the combination of different historical conditions has facilitated the realization that women have to be fully recognized in all democratic institutions.

However, rural areas, especially the ones defined by the agricultural *milieu*, are still ruled by traditional and patriarchal values. To be taken seriously, women have to show both competence and combative spirit. As one woman who inherited a well-established vineyard in Burgundy put it: 'When I took over from my father, they [the other wine-growers from the village] told my husband that the business would collapse'. Such prejudices are still commonplace, and it is only by their visible economic contribution that women can hope to be treated as equal. More than in urban context, women living in rural France still have to prove their status by juggling between different worlds and by acting more as key agents than mediators. The current debate on parity between the sexes in France may offer an opportunity for rural women to challenge traditional views and stereotypes.

The management of these various issues represents a major challenge for the current and future inhabitants of the French countryside. By acknowledging the contribution women can make to these changes, a major step forward will have been achieved. There remains enormous scope for women to participate in both agriculture and rural society more generally. This chapter has attempted to show that women are ready to play a major role, but they can only do so if they are given opportunities and conditions to generate projects and ideas. The political sphere, more than any other, remains crucial if women want to obtain more social visibility. In this area, France is still far behind some of its European partners and the pace of change is very slow, as demonstrated by recent elections. The uniqueness of women's position will be based not only on their role as mediators or agents of change, but also on the development and recognition of their own identity, specific competencies and qualities. They also have to recognize that they are different in a world where differences count. The progress made over recent years has to be built upon and it is likely that, in the future, women will play a major role in restructuring the social fabric and in rejuvenating French political life. Claude du Granrut (2002, p.69), a woman and politician, concluded in a recent publication that women have to play the three following cards in order to engage politically: conviction, generosity and management with distance. Their political engagement needs to enable them to participate fully in public life. If they are given access to more equality, then they could contribute in the way they wish, either as mediators or as agents of change. This could happen at the local level, where new forms of citizenship are encouraged and new forms of rural development are sought. If the results of the last *élections municipales* are repeated, women will be increasingly present in the politics of communes and small towns, and they may play a major role in reshaping of French political life at the local level. Meanwhile, their major contribution remains in bridging the divide between rural and urban parts of France.

References

Bessy-Pietri, P., Hilal, M. and Schmitt, B. (2000) *Recensement de la Population 1999: Evolutions Contrastées du Rural*, Institut National des Statistiques et Etudes Economiques Première 726, Paris.

Braithwaite, M. (1994) *Le Rôle Économique et la Situation des Femmes dans les Sociétés Rurales*, L'Europe verte 1, Office des publications officielle des communautés européennes, Luxembourg.

Braithwaite, M. (1996) Women, equal opportunities and rural development: Equal partners in development, *Leader Magazine*, 5.

Clout, H.D. and Demossier, M. (2003) Introduction, in H.D. Clout and M. Demossier (eds.) *Modern and Contemporary France*, special issue on New Countryside, Old Peasants? Politics, Tradition and Modernity in Rural France, 11(3), pp.259-263.

Confédération Paysanne (2001) *Campagnes Solidaires*, 152 mars. http://confederationpaysanne.fr/cs/152pointvue.htm.

Cross, M-F. (2000) Women and politics, in A. Gregory and U. Tidd (eds.) *Women in Contemporary France*, Berg, Oxford, pp.89-112.

Debatisse, M. (1963) *La Révolution Silencieuse, le Combat des Paysans*, Calman-Lévy, Paris.

Demossier, M. (2000) Women in rural France, in A. Gregory and U. Tidd (eds.) *Women in Contemporary France*, Berg, Oxford, pp.191-212.

Dubois Fresney, L. (2002) *Atlas des Français: Grand Angle sur un Peuple Singulier*, Collection Atlas-Mode, Autrement, Paris.

Du Granrut, C. (2002) *Allez les Femmes! La Parité en Politique*, Descartes et Cie, Paris.

European Commission Directorate General for Agriculture (2000) *Women Active in Rural Development*, Office for Official Publications of the European Communities, Luxembourg.

Eurostat (2002) *Le Guide Statistique de l'Europe: Données 1990-2000*, Office des publications officielle des communautés européennes, Luxembourg.

Ferrand, M. (2002) La place des femmes: Grandes tendances, in *l'Etat de la France. Un Panorama Unique et Complet de la France 2002*, La Découverte, Paris, pp.76-81.

Fischler, F. (1996) Women, equal opportunities and rural development, in a word, *Leader Magazine*, 5.

Gregory, A. (2000) Women in paid work, in A. Gregory and U. Tidd (eds.) *Women in Contemporary France*, Berg, Oxford, pp.21-46.

Gregory, A. and Tidd, U. (2000, eds.) *Women in Contemporary France*, Berg, Oxford.

GREP (Groupe de Recherches pour l'Education et la Prospective) (1998) *Actes du Séminaire Femmes en Milieu Rural: Nouvelles Activités, Nouvelles Qualifications 1995-7 et 1998-2000, Bilan et Perspective*, Europea, Paris.

Herlitzka, H. (1995) *Rapport Final sur la Participation des Femmes dans les Processus Décisionnels du Monde Agricole et Rural*, CEPFAR Centre Européen de Formation pour la Promotion et la Formation en milieu agricole et rural, Bruxelles.

INSEE-INRA (1998) *Les Campagnes et Leurs Villes: Portrait Social*, Institut National des Statistiques et Etudes Economiques, Contours et Caractères, Paris.

INSEE (1993) *Les Agriculteurs*, Institut National des Statistiques et Etudes Economiques, Contours et Caractères, Paris.

INSEE (1995) *Les Femmes: Collection Portrait Social*, Institut National des Statistiques et Etudes Economiques, Service des Droits de la Femme, Paris.

INSEE (2001) *Femmes et Hommes: Regards sur la Parité*, Institut National des Statistiques et Etudes Economiques, Paris.

Jacques-Jouvenot, D. (1997) *Choix du Successeur et Transmission Patrimoniale*, L'Harmattan, Paris.

Jouffe, M. (1996) Women, equal opportunities and rural development: Equal opportunities for men and women: a European ambition, *Leader Magazine*, 5.

Kayser, B. (1990) *La Renaissance Rurale: Sociologie des Campagnes du Monde Contemporain*, Armand Colin, Paris.

Lagarenne, C. and Tabard, N. (1998) *Inégalités Territoriales de Niveau de Vie*, Institut National des Statistiques et Etudes Economiques Première 614, Paris.

Perrot, M. and De la Soudière, M. (1998) La résidence secondaire: Un nouveau mode d'habiter la campagne, *Ruralia*, 2, p.143.

Reynié, D. (2003) Femmes, in P. Perrineau and D. Reynié (eds.) *Dictionnaire du Vote*, Presses Universitaires de France, Paris, pp.449-451.

Rioux, J-P. (2003) Nous n'irons plus au bois, in J-P. Rioux and J-F. Sirinelli (eds.) *La France d'un Siècle à l'Autre*, Hachette, Littératures, Paris, pp.923-931.

Roquette, C. (2003) 10 ans de vacances des Français, in *France: Portrait Social*, Institut National des Statistiques et Etudes Economiques, Paris, pp.159-177.

SEGESA-Service des Droits des Femmes (1995) *Femmes en Milieu Rural: Des Initiatives pour l'Animation, le Développement des Services et l'Emploi*, Ministère des Affaires Sociales, de la Santé et de la Ville, Paris.

Sylvestre, J-P. (2002, ed.) *Agriculteurs, Ruraux et Citadins: Les Mutations des Campagnes Françaises*, CRDP de Bourgogne, Educagri, Dijon.

Toutain, X. (1998) La situation socio-économique des femmes en milieu rural, *Pour*, 158, June, pp.15-25.

Village Magazine (2000) *Women in Rural France*, Hors série, 3 February.

Windebank, J. (2000) Women's unpaid work and leisure, in A. Gregory and U. Tidd (eds.) *Women in Contemporary France*, Berg, Oxford, pp.47-64.

Chapter 4

Rural Women's Employment Opportunities and Constraints: The Norwegian Case[1]

Marit S. Haugen

Introduction

Even though there is a political objective to maintain the residential pattern and to develop sustainable regions in all parts of Norway (*St.melding 31*, 1996-1997), many municipalities have seen their population decline for many years. Urban and densely populated areas continue to gain inhabitants while less central municipalities in the periphery lose people. More young women than men move to urban and densely populated areas, which results in a lopsided population structure with more (single) young men than women in many outlying areas. In order to improve the gender balance in many rural areas, it is recognized that there is a need for improved employment opportunities for women (*St.melding 31*, 1996-1997, pp.14-15). It is further recognized that, in order to succeed in recruiting young families, there have to be job opportunities for both the husband and the wife; the 'two-income family' is seen as the norm. Based on a European research project on rural women, which uses farm women as representatives for rural women, this chapter examines employment opportunities for women in rural areas in Norway. The research question is, what factors influence rural (farm) women's labour situation?

The European project as a whole analyzed the situation of farm women in order to determine what factors promote or impede their participation in the labour market. In exploring this question, we chose to examine five key factors that might influence women's labour situation: 1) external structures, like conditions in the local labour market and local social infrastructure; 2) personal characteristics;

---

[1] This chapter draws on the Norwegian component of the research project: 'Labour situation and strategies of farm women in diversified rural areas of Europe' (Overbeek *et al.*, 1998). This project was funded by the AIR-programme of the European Commission (CT94-2414). The other participants were Greet Overbeek (The Netherlands), Sophia Efstratoglou (Greece) and Elena Saraceno (Italy).

3) household and family structures; 4) farm structures; and, 5) attitudes toward women's role in society. These key factors constitute a set of opportunities and constraints for women's involvement in farm and off-farm work, thereby influencing the economic and social integration of farm women into the rural economy (Efstratoglou *et al.*, 1995). The theoretical point of departure is that women are actors manoeuvring within a gendered society, where their opportunities should be understood within a contextual framework.

Two different study areas, with diversified and less diversified labour markets, were selected to explore the (local) labour market situation of rural women. Both areas are in the county of Nord-Trøndelag in mid-Norway. Based on a context analysis of the study areas, a survey was conducted of 424 women (aged 55 years or younger) in these study areas. In addition, life-story interviews were completed with 20 women. The aim of this chapter is to draw on the data from these sources to illustrate how different factors influence rural women's paid-work situation.

The Study Areas

One of the study areas, the municipality of Stjørdal, has a range of industrial activities as well as one of the county's most active trading centres. Stjørdal offers a wide spectrum of educational opportunities, with more than half the inhabitants (10 500) living in densely populated areas,[2] mainly in the centre of Stjørdalshalsen. According to the 1994 Norwegian Standard Classification of Municipalities, Stjørdalshalsen is an urban settlement at Level 1 (population 5 000-15 000). While half of the inhabitants live in the centre, the rest are spread over six different communities, all within 45 minutes travelling time from Stjørdalshalsen. Stjørdal encompasses Central Norway's main airport and is an important junction for road and trail traffic. Stjørdal is a communication, trading and educational centre in which service industries employ more people than production industries, with primary industries (agriculture, forestry, fishing) collectively employing fewer than manufacturing or construction.

By contrast, all six municipalities[3] in the study area of Namdal have an official centrality Level of 0. In this mountainous, lake-strewn area, agriculture is economically important, with tourism, for salmon fishing, hunting, canoeing and alpine skiing, promoted by a clean natural environment. All except one municipality in Namdal are primary industry areas, with production industries employing more than service industries (albeit manufacturing and construction employ less than primary industries). Containing 9 663 inhabitants, Namdal has a population density of one person/km^2, compared with 19/km^2 in Stjørdal. In Namdal 24 per cent live in densely populated areas, and 76 per cent in sparsely populated areas, with no urban settlement within 45 minutes commuting time.

For these two study areas, recent population changes reveal the same pattern as

---

2   'Densely populated' means that more than 200 people live in an area, with houses not
    more than 50 metres from each other.
3   Lierne, Røyrvik, Snåsa, Namskogan, Grong, Høylandet.

in the rest of the country (Table 4.1); namely, an increase in population in more densely populated areas and a decline in population in already sparsely populated areas, which is due to negative birth rates and out-migration. In Namdal, for instance, all six municipalities are losing population.

Table 4.1 Population change in the study areas of Namdal and Stjørdal

	1980	1997	% change since 1980
Stjørdal	16 107	17 743	10.1
Namdal	10 609	9 663	- 8.9

Source: *Population Census* 1990, *Statistical Yearbook* 1994, *Regionalstatistikk Nord-Trøndelag* 1997.

The composition of the population (sex, age-groups, level of education) is a central issue when discussing labour supply and future developments in rural areas. Here a high proportion of people of retirement age generally indicates an ageing of the population. A higher portion of men than women in age groups where most people establish their own family, likewise indicates a lopsided pattern for the sexes that is associated with out-migration and bachelor problems in many rural areas. A population with a relatively low level of educational attainment is often indicative of a less competitive area economically.

For the two study areas explored here, Stjørdal has a *younger* population, while Namdal has an *older* population compared with the national and county levels (measured by the share of the population aged 67 years or older). When it comes to the balance between women and men (in the age-group 20 to 39 years), there are relatively less young women, compared with men, in the study areas than at the national level. In Stjørdal there are 91 women for every 100 men, while in Namdal the ratio is 88:100. Further indicating a less advantageous labour market situation, the portion of population with compulsory education only is higher in Namdal and lower in Stjørdal than the national average (*Regionalstatistikk Nord Trøndelag*, 1/96, 8/96). Based on information about the inhabitants of these two areas, we can conclude that future prospects seem brighter in the more diversified area of Stjørdal, which has a younger and better educated labour supply, than in the less diversified area of Namdal. However, although existing patterns of development and future potential seem less favourable in Namdal, these conditions might not have a negative effect on farm women's labour opportunities in the short run.

Women and Economic Activity

The rate of economic activity for women has increased tremendously since the beginning of the 1970s. Indicative of this, women accounted for three-quarters of net growth in the Norwegian workforce between 1972 and 1990 (Foss and Tornes, 1992). Women from all generations, ages and life phases have seen increases in the amount of paid-work they undertake. The proportion of women between the ages of 25 and 66 who were full-time housewives fell from 47 per cent in 1972 to 15 per cent in 1987, with this decrease most noticeable amongst younger women. By 1996, the economic activity rate for women in Norway (aged 25-66 years) had reached 76 per cent. This rapid growth in Norwegian women's paid employment has diminished gender differences in labour market participation rates. For comparison, the male economic activity rate in 1996 was 87 per cent (Bø and Lyngstad, 1998), although it should be noted that more women worked part-time than their male counterparts. According to Statistics Norway,[4] in 2002, 43 per cent of women worked part-time compared to 11 per cent of male workers.

Economic Activity by Education Level

Educational levels have increased amongst both women and men in Norway over the last 30 years, but more so for women (*Statistical Yearbook 1994*). Compared with 1980, by 2000 twice as many in the population of 16 years or older had university or college education (22 per cent).[5] Significantly here, there is a positive correlation between level of education and employment rate, both for women and for men, but most notably for women. At the highest level of education, the economic activity rate differs by only 2 per cent between women and men, whilst among women and men with the lowest level of education the difference is 16 per cent in favour of men.

Among the economically active population in the study areas, women's education seems generally to be completed to a lower level than is the case for urban women, but at a higher level than for rural men. In the less economically diversified situation of Namdal the level of education amongst women and men is lower than in the more diversified case of Stjørdal, as well as being lower than county and national figures. Better educated women (and men) are less likely to return to rural home areas, which has the effect of reducing overall education levels for women (and men) in rural areas.

Social Infrastructure

Elements of social infrastructure, like a compulsory school, grocery stores, medical care, post offices and kindergartens, are central institutions for people's everyday lives and well-being. However, most of these institutions require a certain population to be sustainable. The infrastructure that might be of special relevance for farm women's labour situation here are access to formal care facilities and transportation. For formal caring provision levels have certainly been growing since the 1970s. However, informal caring, that is that given by persons who have a family or friendship relation to the care receiver, is still very important. Thus, surveys suggest that half or more of caring activities are still contributed by the

---

4   http://www.ssb.no/emner/06/01/yrkeaku/tab-2003-03-26-04.html.
5   http://www.ssb.no/emner/04/01/utniv/.

informal sector (Kitterød, 1993).

Generally speaking, small municipalities with less than 5 000 inhabitants spend most resources on caring facilities for old people, if this is represented in terms of worker-years of help per person aged over 67 years (Søbye, 1993). In our study areas we found that the public sector (whether for childcare facilities, medical care, elderly care, etc.) is as good or even better in the remote, sparsely populated area (Namdal) as in the central area of Stjørdal. The main difference between these two places was that longer distances had to be travelled to access services in the sparsely populated area. Nevertheless, public welfare provision is as good or even better in the sparsely populated municipalities.

Both public and private kindergartens receive some state subsidies in order to make the level of their charges reasonable for parents. Even so, some kindergartens are quite expensive to use, with charges varying from municipality to municipality. The tax system does not favour both parents working, especially if one of the spouses (most commonly the female) has a poorly paid, part-time job. If parents in addition have to pay quite a lot for childcare, then the extra income they receive from paid-work will have to exceed a certain limit in order economically to justify from a household perspective the decision to use childcare provision (Koren, 1989). From 1 January 1999, parents who do not have a governmentally subsidized childcare arrangement (kindergarten) are entitled to receive from the state a cash benefit (kontantstøtte) for children aged 13-36 months. Three out of four children receive such support (Reppen and Rønning, 1999). This does not necessarily indicate that one of their parents is staying at home looking after children, as parents can arrange private care facilities.

The supply of childcare facilities in our survey areas was such that a lack of childcare facilities did not impose any real constraints on women who would like to work. However, the cost of childcare options can be seen as something of a constraint, as you need a relatively good income in order to be able to afford these services.

While social infrastructure, defined in terms of the supply of care facilities, seems to be quite good in our study areas, compared to the national level, public transportation is poorer the less centralized an area is. Linked to this, the majority of households in our study areas have a car at their disposal. In households with children the figure is as high as 96 per cent, while 72 per cent of households without children have one or more cars at their disposal. It is mainly older people who do not have a car at their disposal.

As regards those who are of working age, then, equality in welfare services has been a political aim, with the result that differences in living conditions between urban and rural areas have more or less vanished.

Government Policies

In order to develop sustainable regions in all parts of the country, with a balanced population pattern and equality in employment and welfare, there is a recognized need to integrate regional (spatial) policy into many (thematic) policy sectors (such as communications, agriculture, fisheries, and the municipal sector). One of the main measures in Norwegian regional policy for this is Statens nærings- og distriktsutviklingsfond[6] (SND – The Norwegian Industrial and Regional Development Fund), which gives loans and grants for industrial development in rural areas. The amounts given depend on where the new activity is to be established. So, in terms of our study areas, Stjørdal is the only municipality in the county of Nord-Trøndelag that is not entitled to support from the SND. Namdal, by contrast, is entitled to all the measures that the SND has at its disposal. Set against this picture, it has to be noted that the location of industry is dependent on many factors other than financial support, such as local infrastructure (such as transport and communication opportunities), know-how, a skilled labour supply, etc. Stjørdal is a 'favourable' location for industry in this respect, while in Namdal the industrial sector is very small.

Gender Policy

Women in Norway have developed a 'dual strategy' towards employment and the family, in which they choose to have both children and paid-work. The most important 'gender' policy that supports this position is probably the building of the welfare state to facilitate parents' economic activity (public care facilities for children, relatively extensive parental leave, and the right to take leave from paid-work in order to care for sick children) and the creation of many new jobs in public services. There has been a political commitment to gender equality set within a framework of social democracy since the 1970s. Thus, in 1978 Norway passed its Law of Equal Rights. In 1986 the Norwegian Government introduced an activity programme for gender equality covering the years 1986-1990, with the explicit aim of gender policy being integrated in all policy-making.

Linked to this, an 'Activity programme for gender equality within agriculture' was established. Since the mid-1980s various kinds of projects have been designed to improve the situation of women in agriculture. One of the tools that came under this rubric was a programme that started in 1986 giving grants to *farm women* who had ideas about establishing their own business. This programme recognized the need for more job opportunities for farm women, so financial and advisory aid was offered to help women implement their ideas.

There are also other kinds of financial support to increase business activity in rural areas. Most important for women are probably the Rural Development Funds (BU-midler), as initiatives from women who want to establish businesses are given priority. Every county manages this fund, and in all counties there should be advisory services for women who would like to establish a business and apply for

---

[6] On 1 January 2004 a new state owned company, Innovation Norway, replaced the following four organizations: The Regional Development Fund (SND), The Norwegian Tourist Board, The Norwegian Trade Council, and the Government Consultative Office for inventors. Innovation Norway promotes industrial development across the nation. It helps release the potential of different districts and regions by contributing toward innovation, internationalization and promotion.

funding support. Yet, in a study of applications and allocations in 1995, it was found that only 34 per cent of the total allocations went to women and only 26 per cent of the total amount of money (Storstad and Haugen, 1997). One explanation for these figures is that women tend to start very small businesses, which often supply enough work for only one person a year or even less. For many women such work is considered a supplementary activity to their other obligations, while for others it is the instigating wish to create their own working place on the farm (or in another location).

The Context Analysis

Despite differences between areas, some recent development trends are similar. In both study areas employment in the service sector has increased the most and this is the most important sector regarding labour demand. Employment opportunities in agriculture are declining in both areas, and there is no signal that implies this decline has reached an end. The decline in the number of farms is however higher in Stjørdal, probably due to its more favourable and competitive labour market, than in Namdal. Regarding labour supply, the economic activity rate for women is equally high in both areas. However, women's employment is very dependent on, and is vulnerable as regards changes in, the private and public sectors. Women setting up their own businesses is relatively few and do not replace the loss of jobs that has occurred within the primary sector.

Advanced state transfer systems have made it possible for local administrations to supply sufficient services in rural areas. Public growth came initially to the education sector and later to the health and social services sector. Public administration at the municipal level was greatly developed after municipal reform in 1964. During the next 10 to 15 years, public services were built up to almost the same level throughout the country (Almås, 1995, 1999). Women especially have gained from this growth in the public sector. In fact welfare expansion through services located at the municipal level led to an equalization of regional differences in women's employment opportunities in the 1970s and the early 1980s.

Table 4.2 Percentage of women in different employment sectors in the study areas, 1980 and 1990

Sector	Stjørdal		Namdal	
	1980	1990	1980	1990
Primary industries	14	7	23	15
Manufacturing industries	12	10	8	7
Public and private services	74	83	69	78
Total	100	100	100	100

Source: *Population Census* 1980, 1990.

Table 4.2 identifies sectoral changes in employment amongst women in the two study areas. There was a significant and similar decline in women working in primary industries in both areas from 1980 to 1990 and a corresponding increase in employment in the service sector. Although the changes were similar, women in Namdal were twice as likely to be active in the primary sector as women in Stjørdal.

Growth in the public sector in the 1970s and 1980s, which gave employment opportunities for women in rural areas, was still important in the 1990s. However, according to *St.melding. 31* (1996-97) there is no reason to believe that the public sector will grow in future. In this context, a rising problem will be how to maintain public services for the inhabitants of small and declining settlements.

All things considered, at present the social infrastructure of the two study areas is fairly equal. As regards employment, although there is a less diversified labour market in Namdal, which indicates fewer job alternatives for women within the area, as agriculture is more important in this area, this might give farm women a better opportunity to work on-farm.

Individual Factors Relevant to the Labour Situation of Farm Women

As regards the individual factors that influence farm women's labour situation, the data this chapter utilizes are drawn primarily from a survey of 424 farm women in the two study areas. The criteria for being selected in the sample surveyed was that woman should live on a farm, either as a farmer herself or as the spouse of a male farmer. Further, she should not be more than 55 years old. The majority of the women in our sample were married or cohabiting, with only 2 per cent single (unmarried, divorced, widowed). The majority have children (95 per cent), with a mean number of 2.6 children each. In comparison, the fertility rate in Norway in 1994 was 1.9 (Noack, 1996). One-third of these farm women have pre-school children (i.e. younger than six years old).

The Labour Situation of Farm Women

Farm women make a variety of significant contributions to the household and the farm business. The majority of farm women are involved in farming and in nearly all types of (on-farm or off-farm) work, but to varying degrees. Even though many women have an off-farm job, they tend to be versatile and flexible, contributing to farm work when necessary. Sixty-eight per cent of the women have an off-farm job, and nearly one-fourth are involved in other gainful economic activity on the farm. This includes farm-based activities like farm tourism, camping, producing home crafts, processing farm products and the direct sale of farm products. Accompanying this work, farm women still have the main responsibility for household work. One-third of interviewed women spend 32 hours or more a week on household and caring activities. Twenty-five per cent are engaged in voluntary work. Hence many farm women combine different types of work activities.

To simplify the analysis of work activities, I have classified farm women into four *main* working categories with regard to their labour situation (based on the type of work undertaken and its intensity).[7] The distribution of farm women into these four groups, according to their main activity pattern, shows that the largest group of farm women work mainly off-farm (38 per cent). The second largest group work mainly on-farm (30 per cent), while 23 per cent are pluriactive, in that they work both on and off the farm. Only 9 per cent are classified as mainly housewives.

The majority of farm women who work on the farm describe their position on the farm as partners, rather than as assistants. Even amongst those who work mainly off-farm, and amongst housewives, more women consider themselves to be partners in a farm enterprise than view their farm role in any other way (Table 4.3). This might indicate a greater awareness of the importance of women's contribution to the family farm as a business, together with a shared interest in, and responsibility for, the farm with their husband or partner. Yet few women saw themselves as the head of a farm enterprise (in total 5 per cent).

Table 4.3 Percentage of farm women in different labour situations and perceived role on farm by labour situation

	Mainly on-farm (N=127)	Mainly off-farm (N=159)	Pluriactive (N=98)	Mainly house-wives (N=40)	Total (N=424)
Labour situation	30	38	23	9	100
Position on farm *					
Head	9	2	7	2	5
Partner	73	44	76	40	60
Assistant	18	32	17	29	23
Not involved	-	22	-	29	12
Total	100	100	100	100	100

Note: * As perceived by the women themselves.
Source: Haugen and Blekesaune (1996).

It should be noted that there are significant differences in farm work intensity across the groups identified above. Women who work mainly on-farm tend to work most regularly full-time on the farm. The pluriactive group allocate their working hours between farm and off-farm work. By contrast, the majority of women who worked mainly off-farm shared with those who were mainly housewives in having a tendency to work only occasionally or seasonally on the farm. These two groups

---

[7]   This classification was developed in the EU-project 'Labour situation and strategies of farm women in diversified rural areas of Europe' (Overbeek *et al.*, 1998).

are a flexible labour force that is involved with farm work when necessary.

Comparing women's farm work intensity with that of their partners or spouses, we find that more men (70 per cent) work regularly full-time on-farm. The reason why so many women often work part-time (on-farm or off-farm) is that they have the main (or even sole) responsibility for domestic and care work.

Off-Farm Work

Women with off-farm jobs commonly have a regular all-year, part-time job. Amongst those working mainly off-farm, 43 per cent work full-time (i.e. 32 hours a week or more), whereas only 19 per cent in the pluriactive group works full-time off-farm. Nine out of 10 women with off-farm jobs are employees. The others are self-employed or work in a family business. The majority work in the public sector (health, care services or teaching), which underlines the importance of the welfare state (and of regional policy) in creating and maintaining jobs and services in sparsely populated regions. Similar patterns have been noted elsewhere in Europe, as with Spain and Sweden; e.g. Persson and Westholm, 1994; Navarro, 1999). Compared with the mainly off-farm women's group, the pluriactive group have slightly inferior labour market positions (e.g. more seasonal work).

Farm Women's Income and Social Security

Farm households have many potential income sources, from farming, farm-based activities, off-farm work, pensions and social security. Altogether 54 per cent of the farms in the sample have their main income from farming, while 44 per cent earn it from off-farm work. 'Other' farm-based activities are the most important income source for a small number of farms (1 per cent). Nearly half the farms are low-income farms. The data show that only about half of farm women have a total farm income above NOK 160 000 (20 000 Euros).[8]

Since 1987 spouses (read women) with farm work have had equal rights as regards their own farm income (no upper limit) and pension entitlements (*Ot.prop. 12*, 1986/1987). However, a relatively large proportion of women who are actively involved with farm work have no farm earnings. Thus, 22 per cent of the mainly on-farm group, and 37 per cent of the pluriactive group, have no income from farming. In this context, the off-farm labour market generally offers farm women much better incomes and pension entitlements than the farm sector (Table 4.4).

As the Norwegian welfare state is built upon individual rights, in many cases the kind and level of social security rights available to an individual depends upon a person's taxable income. Farm women's lack of formal income, or their low earnings, implies that they will receive only a basic pension disbursement at retirement age and that their social security rights are not as good as they would be if they were employed and had a higher formal income (which then impacts on

---

[8]   The Norwegian Income Distribution Survey shows that the average 1995 wage income for men was NOK 175 013 (22 000 Euros), and NOK 110 330 for women. The average income for the self-employed was slightly lower.

payments related to disability, sickness, unemployment and pensions). Accordingly, women working on-farm were not satisfied with their social security rights. However, three-quarters of those working mainly on-farm compensated for their lesser formal rights by buying private insurance. A simple question arises, which is why some women work on-farm while others choose better-paid off-farm jobs? As the analysis that follows shows, there is no simple answer to this question.

Table 4.4 Percentage of women by farm and off-farm income categories

Women's income in NOK (Euros):	Women working mainly on-farm	Women working mainly off-farm
No income	22	0
Less than 80 000 (< 10 000 Euros)	28	16
80 000 - 159 000 (10 – 20 000 Euros)	36	41
160 000 and more (≥ 20 000 Euros)	14	43
Total	100	100
(Number of informants)	(127)	(159)

Source: Haugen and Blekesaune (1996).

Individual Factors and Household Structures

Women's educational levels are found to be an extremely important determinant of their labour strategies and their opportunities within the labour market. In this regard it should be noted that there are significant differences in the educational level of farm women in different labour market positions. Women working mainly on-farm have a significantly lower educational level than women working off-farm (31 per cent of women working mainly on-farm had only compulsory education compared with 18 per cent working mainly off-farm). This point applies not just to schooling but also to vocational training. A little more than half of all farm women had some form of vocational training. But women with off-farm work in general had a significantly higher rate of vocational training (70 per cent) than women in the other groups (43 per cent). The higher the level of formal education and vocational training, the more likely a woman has an off-farm job. The fact that women with higher education tend to have off-farm jobs indicates that women with the best opportunities to take advantage of labour market possibilities choose an off-farm job rather than farm work. A low level of education and a lack of vocational training might be a real constraint for women who want off-farm work.

The majority of all farm women have experienced the external labour market, but as many as half who today are mainly employed on the farm have no outside work experience, or have been out of the off-farm labour market for at least 10 years. Among mainly the housewives group, one-quarter have never been in the labour market, or it is 10 years or more since they were in paid employment. Due to changes in nearly all sectors of employment, these women will probably have little chance of (re)entering the labour market, especially without further training.

As regards other potential restrictions, having a driving licence is a must for mobility reasons in most rural areas. Distances are large and women without a driving licence and the use of a car have limited opportunities to take off-farm jobs, or their choices are restricted by transportation difficulties. However, practically all of the women (98 per cent) in the sample hold a driving licence and most women working off-farm have a car available for their journey to work. But women who were not working off-farm reported that they were less likely to have a car at their disposal if they wanted one. Whether this is an actual constraint might be questioned, as it is probable that families will purchase a car, or a second car, if one or both partners need it for work.

It might be supposed that having pre-school children was an important constraint on women's job choices. However, in Norway the majority of mothers now continue to work after the birth of their first child (Ellingsæter and Rubery, 1997). Thus, from 1980 to 1990 the labour force participation rate of mothers with a youngest child under three years old grew from 47 to 69 per cent, and among mothers whose youngest child was three to six years old from 57 to 74 per cent (Kjelstad, 1991). Not surprisingly, then, we found no significant differences in the likelihood that women in the four labour groups had pre-school children. As the majority of women live in nuclear families with only one adult woman present, we have no evidence that family type makes any difference to women's employment situation.

Farm Structure (Labour Demand On-Farm)

Women are most involved in farm work on dairy farms and less involved in grain-producing farms. This supports the hypothesis that women are a flexible labour force on the farm; if farm production is intensive, as dairy production is, then women tend to work more on-farm. However, it might also be the other way round, in that farm production might become more extensive as women seek off-farm employment. Exploration of this possibility would necessitate an examination of the timing and decision-making process associated with farm production changes and the take-up of jobs off-farm.

For the present what can be said is that the relationship between farm structure and women's work indicates that women are more involved in labour intensive enterprises. It can also be noted that the farm needs a certain production size in order to give labour opportunities for women. If the farm has a production equivalent of 1.5 worker-years or more, then women tend to increase their labour participation on-farm. This means that women are most involved on relatively large or on medium-sized family farms.

Labour Market Structures and Rural Context

It might be assumed that labour opportunities are poorer in the less economically diversified area of Namdal than they are in Stjørdal. However, the findings of the survey do not support this, even though there are some differences between the two places. In the economically diversified area, farm women tend to work *either* mainly off-farm *or* mainly on-farm, with only 16 per cent as pluriactive. In the less diversified area a higher proportion of farm women combine farm and off-farm work (28 per cent). In total, 59 per cent of women in the diversified area and 62 per cent in the less diversified area have an off-farm job.

Although there is only a minor difference between areas in women's opportunities to find an off-farm job, women in the less diversified area of Namdal are more likely to combine their off-farm job with farm work. The majority of farm women in both areas were employed within the public service sector; whether this be in health care and social work, in teaching or in administrative and clerical work. The distances that had to be travelled to reach present off-farm jobs were roughly equivalent in the two areas, if measured in terms of actual journey times. Hence, the majority spent less than half an hour travelling to their job in both areas. This indicates that both study areas have a developed social infrastructure, which supports farm women's opportunities and participation in the labour market. Illustrative of this, in Namdal nearly three out of four farm women with pre-school children use kindergartens, even if mainly on a part-time basis, while women in Stjørdal are less likely to use any childcare facility, so less than half use kindergartens.

In both areas 3 per cent of farm women are registered as unemployed. However, many women who want a job do not register themselves as unemployed, as they have no rights to unemployment benefits if they have not had taxable income during the previous 12 months. In line with this more women than those who are registered as unemployed have applied for a job the last 12 months, with more having done so in Namdal (9 per cent) than in Stjørdal (5 per cent). And there are even more women looking for information about vacant jobs, with 16 per cent reporting to have done so in Namdal and 8 per cent in Stjørdal. This indicates that there are women in both areas who are looking for new opportunities in the labour market, with more doing so in the less economically diversified area of Namdal than in Stjørdal. There are various reasons for looking for a job – such as unemployment, underemployment or merely a wish to change a present job – but the survey results show that women in Namdal are less satisfied with their present job situation than women in Stjørdal. One reason for this is that there is a more limited breadth to the openings that are available in Namdal, with women more likely to find that they are 'over-qualified' for available posts and, as a result, view available jobs as qualitatively inferior.

Farm Women's Motivation to Work On-Farm and Off-Farm

Life-story interviews with a sample of 20 farm women in different labour positions, along with questionnaire survey, revealed a strong work-orientation amongst farm women. Looking at values regarding women's work, nearly all (93 per cent) survey respondents agreed with the statement that: 'It is important for women to have their own income', while 98 per cent agreed with the statement that 'It is equally important for women as for men to have paid work', and 97 per cent agreed that 'It is important for women to realize their own job aspirations'. We can conclude that farm women express an egalitarian gender ideology regarding the importance of (paid) work for women. Work is an important part of their identity, whether they choose to work on the farm or off-farm or as some combination of the two. Yet this work orientation does not exclude a family orientation, because most farm women, like other Norwegian women, combine work and family life in a 'dual strategy'. But women's motives for work are complex and they are hardly possible to grasp through a questionnaire survey. This is because questions about motivation are difficult to draw out using a standardized set of questions. People tend to rationalize and legitimize their choices according to what they perceive to be a commonly accepted rationale. A family orientation might be a socially more acceptable reason than an individualistic orientation.

That noted, the most important motive that was given for working on the farm for women who work mainly on-farm, and for those who are pluriactive, was 'to contribute to the family income'. This indicates that women look upon themselves as joint breadwinners with their husbands or partners, having a joint responsibility for family welfare. The workload on the farm that demands a joint contribution is also important, as well as building up and continuing the family farm business. More individual motives, such as pleasure and interest in work, are also important, but less important than collective motives. Here it should be noted that there are no significant differences in motivation between women who work mainly on-farm and the pluriactive group, nor is there a difference between older and younger women.

Women's motivations for working off-farm are dominated by the same motives as those for women working on-farm; with 'to contribute to the family income' as the most important reason given. However, individual motives, like having one's own income and building up social security rights, are also important reasons for working off-farm. It is worth noticing here that pluriactive women put much more emphasis on building up social security rights, which they cannot do so well in agriculture. The social motive to get in touch with other people is also important for all women working off-farm. The motives 'to use my education' and 'to have my own income' are significantly more important for women working mainly off-farm. Such women seem to be more individually career-oriented than pluriactive women.

Farm Women's Labour Strategies and Preferences for Change

In so far as the women had specific work aspirations when they were young, these tended to be within traditional 'female' areas, like 'working with people', as exemplified in jobs within care sectors, in health, in teaching or in clerical work. However unintentionally, the fact that the majority chose a typically 'female education' seems to match well with the labour market situation of many rural areas. But there are some differences between older and younger women; not so much in (occupational) aspirations as in the realization of their aspirations. This is as a result of better opportunities and changed attitudes, which mean that younger women are more likely to have completed vocational training, and to have been able to realize their aspirations.

That said, the majority of farm women are satisfied with their present labour situation and do not aspire to any major change (Table 4.5). Women who are mainly working off-farm seem most satisfied with their present labour situation. Only 10 per cent want to quit their off-farm job, and eventually work more on-farm. More of those working mainly on-farm, and those who are mainly housewives, want to see more substantial changes.

Twenty per cent of the women working mainly on-farm and 39 per cent of those who are housewives want an off-farm job. It is interesting to note that as many as one-third of the housewives group would like to work more on-farm. This illustrates that (limited) labour demand on-farm might be a constraint on women's work. Among the pluriactive group some want to work more on-farm, while others want to reduce their farm work or quit their off-farm job.

Table 4.5 Percentage of farm women by present labour situation and change wanted to labour situation

	Mainly on-farm (N=127)	Mainly off-farm (N=159)	Pluriactive (N=98)	Mainly house-wives (N=40)	Total (N=424)
No change wanted	65	67	50	42	60
Change wanted on-farm*					
Want more work on-farm	14	30	30	33	26
Want less work on-farm	10	3	13	6	8
Change wanted off-farm*					
Want an off-farm job	20	-	-	39	10
Want to quit off-farm job		10	16	-	8

Note: * These are not mutually exclusive. A woman might want both on-farm and off-farm changes.

Source: Haugen and Blekesaune (1996).

Farm Women's Future Aspirations, Expectations and Initiatives

Agricultural decline implies fewer future opportunities for work in the sector. This is in accordance with what farm women themselves expect; for more than half expect income prospects in farming will be worse in the next five years. There is, however, one trend that might change this picture, which is a growing interest in ecological farming and rising demand for ecological products. Ecological farming is more labour intensive than traditional farming. Moreover, agricultural policy is increasing financial support for ecological farms (Landbruksdepartentet, 1998). In our survey, 6 per cent of farm women in the two study areas expected to change

toward ecological production on their farms within the next five years.
Interestingly, more women working mainly on-farm expected this change.

As well as asking about future farming conditions, women were also asked
about other work possibilities, such as whether they had searched for information
about available jobs and whether they had applied for a job the last 12 months.
They were also asked about how they perceived labour opportunities in their area.
Nearly half the farm women expect that they will start to work or increase their

**Table 4.6 Percentage of farm women by aspirations, expectations and active
searches to change labour situation**

	Mainly on-farm (N=127)	Mainly off-farm (N=159)	Pluri-active (N=98)	Mainly house-wives (N=40)	Total (N=424)
Perceptions					
Will work (more) off-farm in 5 years	31	48	56	56	46
Poor local job prospects	57	38	48	63	48
Changes wanted					
Want an off-farm job	20			39	10
Want to change off-farm job		14	11		8
Want to start a new activity on-farm	10	14	19	19	15
Initiatives					
Recent further education/training	9	27	24	25	21
Information search on available jobs	13	9	10	23	12
Applied for a job the last 12 months	4	6	9	15	7
Registered as unemployed	5			8	3

Source: Haugen and Blekesaune (1996).

work commitments off-farm within the next five years (Table 4.6). The majority of
women who do not have an off-farm job perceive in general that job opportunities
in their area are poor. Many doubt they will be able to find paid-work. This figure
indicates growing pressure on the labour market (from the supply side), as there are
many farm women who, for various reasons, will enter the labour market or
increase their off-farm work. Some of these women are actively searching for a
job, while others seem to be a more of a latent labour force. Women classified as
mainly housewives are most likely to be active in searching for a job. But it is
interesting to note that women who already have an off-farm job seem to be active
in searching for another job. This might indicate that they are not satisfied with
their present job or it might indicate a general mobility between jobs. It is
noteworthy that many who are actually searching for a job *are not formally
registered as unemployed*. The reasons given for this, amongst others, are that they
are not entitled to unemployment benefit if they have not been in the labour market
in recent years, or more commonly 'it is useless' as it will not create more jobs, or
they know there are no jobs. The 'real' unemployment rate is therefore somewhat
higher than official statistics indicate.

One way farm women might improve their job opportunities is to start a new business activity. A total of 15 per cent of farm women would like to start a new (business) activity on their farm. Pluriactive women and housewives seem to be most enthusiastic about such an idea. These pluriactive women are inclined to prefer to create a labour situation where they can work more on the farm and quit their off-farm job, while housewives report that starting a new activity on the farm is their main (sole) opportunity for getting a job.

Active Women: A Synthesis

As regards the main results from this research, it is pertinent to note that farm women in general have a high economic activity rate, with more than 90 per cent working either in agriculture or in off-farm jobs. The women's levels of education and vocational training are the most important factors determining their labour market position. The higher the level of education, the more likely a woman is to have an off-farm job. In this regard, it should be noted that women who work mainly on-farm are not only more likely to have a farm background than those who do not work on the farm, but to also have a significantly lower educational level than women working off-farm. Nearly one-third have completed only compulsory education. Quite apart from their own background, another factor encouraging on-farm work is the nature of the farm enterprise. Farm women are most likely to be involved in farm work on intensive (dairy) farms, but whether farm production becomes more extensive when women secure off-farm employment, our survey cannot answer. What can be stated is that our life-story interviews showed that younger women were more likely to continue off-farm work after marrying a farmer. In general terms, though, many women reported that they wanted to involve themselves more in farm work. However, those working mainly on-farm have less taxable income and thereby less social security rights than those who work off-farm. This means that agriculture cannot compete with the labour market in terms of formal rights, which might be one reason why so many women choose or aspire to have an off-farm job.

When seeking such off-farm opportunities, quite surprisingly, the labour markets in the more and less diversified areas seem to offer more or less the same job opportunities for women. However, there is a tendency for women in the less diversified area to work part-time and to combine an off-farm job with farm work to a greater degree than women in the more economically diversified area. Even so, it is noteworthy that off-farm employment is close in the two areas investigated. An important reason for this is the nature of public services provision in rural Norway, for the majority of farm women in both areas are employed within the public services.

In seeking to understand the factors that promote or impede participation by farm women in the labour market, four labour groups were identified, which differed according to women's main activity (eight hours or more per week). This analysis has revealed different strengths and weaknesses in existing labour market situations, and in relation to farm women's aspirations regarding their work situation. In this regard, one-fifth of the women who work mainly on-farm would like to have an off-farm job. Controlling for education we found that more women with only compulsory school would like to have an off-farm job, than among women with secondary or higher education. Those with better education might have chosen farm work more actively, while women with lower educational levels might have felt it was their only (or best) alternative. Amongst younger women there is a difference between those who have pre-school children or do not. More women with no pre-school children at home would like to have an off-farm job.

There is also a difference between younger (under 40 years old) and older farm women, as more older women would like an off-farm job than their younger counterparts. There is a group of older women with low levels of education who would like to have an off-farm job, but who have difficulties realizing their aspiration. This indicates that younger women have more opportunities to choose their present labour situation (either on-farm or off-farm), while older women have experienced more constraints. Our qualitative interviews support this picture, for older women's experiences when they were younger provided less incentive to pursue higher education or vocational training. When they married they were more likely to give up their (unskilled) job in order to work on the farm and in the household. Then, with changes both in farming and in their family (e.g. as children grew up), these older women have more time available. However, changes in the labour market have raised the demand for skilled and experienced workers, which restricts the possibilities of these women taking up job opportunities. As one woman of 49 told us during interview: 'If I knew 15 years ago what I know today, I would have kept my off-farm job'. Today she finds that (re)entry into the labour market is nearly impossible.

At an individual level the weakness amongst women who would like an off-farm job is their relatively low level of education, a lack of vocational training, and dated or no experience in the labour market. Their strength is their mobility, as they have a driving licence and commonly have a car available. But the majority of women working mainly on-farm are living on dairy farms. Here the workload on the farm demands a work contribution from them, and this is a constraint for those who would like to start an off-farm job.

Even so, women working mainly on-farm perceive job opportunities in their local area as poor (with no difference between the two study areas). Yet only 11 per cent say that the most important reason for not having an off-farm job is lack of available paid-work. During the qualitative interviews it became clear that many farm women do not want a low skilled, physically demanding job (like cleaning), as they indicate that 'they have enough hard work at home'. There might be job opportunities, but not sufficient to find a preferred job.

It is women with mainly off-farm work who have the highest educational level among farm women. The majority of these women are employees, who work regularly full-time or have worked for long periods in a part-time capacity. They are most likely to have a labour situation in accordance with what they aspired to when they were 20 years old. Their income is an essential contribution to the farm household, to the extent that on average they contribute nearly half of the total farm household income. These women who work mainly off-farm are most likely to consider themselves economically independent.

Even so, there is a tendency for younger and better educated women to want to work more on-farm. However, we do not know exactly why they say this – whether it is a preference for farm work or it arises more from a greater need for their labour input on the farm than they are able to fulfil. In both cases the labour demand and economy of farming might be constraints. Another explanation is that some young women find farm work easier to combine with childcare. Then there is the case of women who want to change their off-farm job, which might indicate that there is a lack of suitable jobs corresponding to their qualifications and aspirations (e.g. Little, 1994), but just as likely is a wish to try something else, just to have a change.

Half of the women who combine work on-farm and off-farm do not want to change their present labour market situation. However, one-fifth would like to work more on-farm in addition to their off-farm job and some would quit their off-farm job in order to work more on-farm. Some of the women would prefer to work less on the farm, or quit their off-farm job, to reduce their total workload. However, the main finding is that pluriactive women would rather increase than reduce their labour input on the farm.

Our survey and life-course interviews with farm women showed that a permanent housewife role was considered a lifelong strategy by hardly any of the interviewed women. Mainly housewives are most likely to report that their present labour situation is not as they wanted it to be when they were 20 years old. The group of housewives is however a heterogeneous mixture of women. Among these women there are those who temporarily give priority to their present care responsibilities (whether pre-school children or others), those who might have health problems which prevent them from being economically active, and those women who unintentionally are under-employed. An example of the last of these is the (low intensity production) farm that does not generate enough farm work, and from which women would like to have an off-farm job. Here lack of required qualifications and age are most commonly mentioned as individual constraints on attaining this goal. Set against this, the strength of these women is their mobility; they have a driving licence and commonly a car available.

At a farm level there seem to be few constraints on housewives having an off-farm job. This is because the farm cannot supply enough work for women for farms to offer a serious job alternative. But even given this situation, at the household level many women mention family motives for not working off-farm. Illustrative of the argument put forward is the statement that: 'it is best for the children that I stay home'. These women commonly aspire to work more when their children grow up, when they expect they will be looking for an off-farm job. Yet this set of expectations is set within a context in which the group of mainly housewives perceive job opportunities in their area as the worst of any of the four women's groups. In reality a lack of vocational training makes the employment opportunities of these women less favourable than for others.

Policy Considerations

Government policy in Norway since the 1980s has had 'the woman question' high on the agenda, with the aim of engagement in issues in order to gain more gender equality (regarding rights and opportunities). Today, this 'woman question' has become integrated into most government policy sectors. However, women are not a homogeneous group. Women do not necessarily have common interests and aspirations. There are areas in women's lives that are of major importance for their opportunities to realize their own aspirations, but where policy can scarcely interfere, as these realms are considered to belong to the personal sphere (e.g. women's domestic roles, motherhood, the division of labour within the household and decision-making processes amongst spouses, etc.).

The main finding from this study in Norway is that the majority of farm women are satisfied with their present labour market situation. A general conclusion is that farm women want to be economically active, either on-farm or off-farm. There is a group of women who want an off-farm job, but because they *lack relevant qualifications* (either vocational training or relevant labour market experience), and partly because there is a *lack of suitable jobs*, they do not succeed in realizing their aspirations. Some women would like to work more within agriculture, but lack of labour demand on the farm and poor income prospects in farming are real constraints. To start another form of gainful activity, other than becoming an employee, is considered an interesting opportunity, but is regarded as less likely as a viable income-source. This study has shown that many women enjoy farm work and want to work (more) in agriculture. However, today women (as farm partners and assistants) have less status, recognition and income (if any) from farming, compared with men, and there is a mismatch between women's labour input and their income. This means that farm women's social security is not as good as it would be if they held an off-farm job. Thus, those with farm work have only a basic pension when they reach retirement age, they also have less rights to maternity leave, lower unemployment benefits, inferior sickness benefits, and poorer disability allowances. Inferior working conditions should be seen as something that is of major importance for the sector itself.

For the future it needs to be recognized that the relatively small farms that characterize Norway (and the study areas) commonly do not generate sufficient work and income for two persons. Only more labour intensive modes of farming – like ecological production – might change this picture. More information, advisory services and training courses in this type of farming, along with better access to the processing and marketing of products, could be a step in this direction. However, this will not change the main picture; which is that the majority of farm women want to have (and actually do have) an off-farm job, eventually in combination with on-farm work.

Added to which, young women who marry male farmers today most likely have education, professional experience and career plans. They invest time and money in education, indicating that education and paid-work are part of a larger life plan;

with marriage and having children generally not leading to a change in these women's work aspirations. Together with the fact that the majority of farm families depend on an off-farm income, this implies that job opportunities for women are increasingly important in order to maintain the population of many rural areas. In this regard public sector expenditure is extremely important, as most service jobs that women hold are in the public sector.

Adult Education and Training Opportunities

It is the case that some farm women want to change (improve) their labour market situation by getting an off-farm job. The main constraint on them achieving this, in addition to the lack of suitable jobs, seems to be a lack of relevant qualifications and labour market experience. Better opportunities and access to adult education for those farm women who need more qualifications in order to get a job would be a mechanism for enhancing prospects of employability. But for this to yield fruit, education and training programmes need to be adapted to local labour needs. In order to overcome distances and reach women in remote areas, courses could partly be taught by distance teaching, using modern telecommunications.

In the study areas there are several organized AMO courses (labour-market oriented courses), which are offered to people who are seeking a job and are registered as unemployed. People above 25 years old are given priority on these courses. The intention is to prepare and qualify participants so they can find a job. Various AMO courses are adapted to local labour market situations, like those focusing on 'care-work' and 'tourism', while 'secondary education' is offered to those who need this in order to take a place on a higher education course. During the time they participate in an AMO course, participants receive financial support for living and travel expenses, a family allowance, and funds to cover the costs of childcare. Significantly in the context of this chapter, 58 per cent of participants on these courses are women.

However, many farm women who are not registered as unemployed do not have access to these AMO courses. Women who are job searchers need to register themselves as unemployed, as they then have better access to training courses, to job information, etc. But it would also be advantageous if courses were arranged for farm women about starting their own business (viz. 'other' farm-based activities). It is worth noting in this regard that 88 per cent of those who expressed the view that they would like to start a new business activity indicated that they need advice and information on other women's experiences in order to know how to start the process of business formation. This could be organized in AMO courses on 'How to establish your own business', 'Marketing', etc.

Increasing Know-How About Information Technology in Rural Areas

A further mechanism that could be called on to improve the work possibilities of farm women is to increase their know-how about information technology. This could increase women's opportunities in the labour market (including possibilities for 'long-distance' work and using a home office). Farm tourism is an example of an activity that could be facilitated by the use of the internet (marketing, booking etc.). Yet there is an increasing need for the know-how and the ability to use information technology in our society. More and more routine tasks, like paying bills, transferring money, etc., could be undertaken more effectively using a personal computer. Further access to the information flow that is today available on the internet could also be a major source of information that could widen work possibilities. Indeed, many of the services that are available on the internet could be more important for rural people in remote areas, where distances to banks, libraries, etc., are greater.

It is worth noting here that, in Sweden as in the UK, they have started 'tele-cottages' in some rural areas (e.g. Clark, 2000). This idea could be adopted in Norway. Not only would this provide an information technology facility in remoter places but it could also be a meeting place where women (and others) could become more familiar with new opportunities, as well as gaining the ability to use, acquire and update knowledge through courses, the exchange of experience and advice regarding programmes, investments, etc.

Networks Among Rural Women

In the survey reported in this chapter it was found that many women who would like to start a new business stated that they would learn from other women's experience. A way forward in enabling a greater exchange of experiences, and a mechanism for improving the flow of advice to farm women, would be for them to participate in national or more limited networks with other women. Most business networks and organizations in rural areas tend to have a 'men only' membership, so women need to be taken in or to establish their own networks or organizations (Almås, 1995, 1999).

In Norway today there is increasing interest in co-operative initiatives and in new businesses within agriculture. In 1991 the Ministry of Agriculture appointed a national committee whose main objective was to encourage more co-operation among farmers. In connection with this, advice and economic support have been made available for agriculturally-related co-operative initiatives. One of the results of the work of the national committee has been the establishment and diffusion of Rural Services in Norway. Rural Services are businesses that are established by farmers themselves in order to organize the common marketing of their services. Here services are provided outside farming to municipalities, businesses and private individuals. The main idea behind the establishment of Rural Services is that members can make use of their 'spare' labour, of their competencies and of their equipment (machinery), in order to increase their earning power. At the same time farmers retain their autonomy, as they are still self-employed. Rural Services accepts many different types of tasks, with only imagination and competences setting the limits. As such, Rural Services represents a positive element and service resource for the local community and its inhabitants. For those who are part of Rural Services it represents a flexible and alternative job opportunity. In Sweden women have started their own businesses (in a Rural Services type of format) and there should be potential for women in Norway to do the same. For the future it looks likely that there will be increasing demand for private services that are versatile and flexible, which the public service sector cannot meet. Many people will have the means to pay for these services, as private income in rural Norway is increasing and is not far behind the level of urban income.

So, while there is still scope for improvement, it should be recognized that the emphasis in rural policy on creating new employment opportunities for women has had a positive effect. Thus, Pettersen and colleagues (2000) have evaluated the installation grants that were given to women who wanted to establish their own business, and have found that 60 per cent of women who received a grant in 1994 had established a business that was in operation five years later. The businesses established are small, for on average they only employ one or two persons. However, more than half of the women who ran these businesses expected their business to grow within the next five years. Furthermore, the businesses that women established gave them what they wanted; namely, an interesting and challenging job. Extending this prospect to more rural women is a task for the future.

References

Almås, R. (1995) *Bygdeutvikling*, Det Norske Samlaget, Oslo.

Almås, R. (1999) *Rural Development – a Norwegian Perspective*, Centre for Rural Research Report 9, Trondheim.

Bø, T.P. and Lyngstad, J. (1998) *Arbeid, Sosialt utsyn*, Statistics Norway, Oslo-Kongsvinger, pp.83-98.

Clark, M.A. (2000) *Teleworking in the Countryside*, Ashgate, Aldershot.

Efstratoglou, S., Efstratoglou, A. and Mavridou, S. (1995) *Theoretical Aspects of the Methodological Design of the Research Programme: Theories on Farm Women's Participation in Labour Markets and Relevant Hypotheses*, Internal Report, Agricultural University, Athens.

Ellingsæter, A.L. and Rubery, J. (1997) Gender relations and the Norwegian labour market model, in J.E. Dølvik and A.H. Steen (eds.) *Making Solidarity Work? The Norwegian Labour Market Model in Transition*, Scandinavian University Press, Oslo, pp.111-154.

Foss, O. and Tornes, K. (1992) *Arbeidsmarkedet og velferdsstaten: Viktige rammebetingelser for kvinnerettet regionalpolitikk for 1900-tallet*, Norwegian Institute for Urban and Rural Research (NIBR) Notat 108, Oslo.

Gjerdåker, B. (1995) *Bygdesamfunn i omvelting 1945-1996*, Landbruksforlaget, Oslo.

Haugen, M.S. and Blekesaune, A. (1996) *The Labour Situation of Farm Women in Norway*, Centre for Rural Research SFB Report 4/96, Trondheim.

Kitterød, R.H. (1993) Uformell omsorg for eldre og funksjonshemmede, *Sosialt Utsyn 1993*, Statistics Norway, Oslo-Kongsvinger, pp.403-413.

Kjelstad, R. (1991) 1980-årene: Småbarnsmødrenes tiår på arbeidsmarkedet. *Samfunnsspeilet*, 5, pp.16-19.

Koren, C. (1989) Lønner det seg for mor å jobbe? *Samfunnsspeilet* 1/89, Statistics Norway, Oslo-Kongsvinger, pp.26-29.

Landbruksdepartementet (1998) *Handlingsplan for økologisk landbruk (1998-1999)*, Revidert rapport, Landbruksdepartementet, Oslo.

Little J. (1994) Gender relations and the rural labour process, in S.J. Whatmore, T.K. Marsden and P.D. Lowe (eds.) *Gender and Rurality*, David Fulton, London, pp.11-30.

Navarro, C.J. (1999) Women and social mobility in rural Spain, *Sociologia Ruralis*, 39, pp.222-235.

Noack, T. (1996) Familieutvikling i demografisk perspektiv, in B. Brandth and K. Moxnes (eds.) *Familie for tiden. Stabilitet og forandring*, Tano Aschehoug, Oslo, pp.11-29.

Ot.prop 12: Endringer i skatteloven (1986/1987), Finans- og tolldepartementet, Oslo.

Overbeek, G., Efstratoglou, S., Haugen, M.S. and Saraceno, E. (1998) *Labour Situation and Strategies of Farm Women in Diversified Rural Areas of Europe*, Office for Official Publications of the European Communities, Luxembourg.

Persson, L.O. and Westholm, E. (1994) *Europas landsbygd i förändring*, Expertgruppen för forskning om regional utveckling ERU Report 83, Fritzes Kundtjänst, Stockholm.

Pettersen, L.T., Alsos, G.A., Anvik, C.H. and Ljunggren, E. (2000) *Sammendragsrapport: Blir det arbeidsplasser av dette da, jenter? Evaluering av kvinnesasingen i distriktspolitikken*, Nordlandsforskning Report 3, Bodø.

Population Census 1980, 1990, Statistics Norway, Oslo-Kongsvinger.

Regionalstatistikk Nord-Trøndelag. 1/96, 8/96, 1-2-3/97, Statistics Norway, Oslo-Kongsvinger.

Reppen, H.K. and Rønning, E. (1999) *Barnefamiliers tilsynsordninger, yrkesdeltakelse og bruk av kontantstøtte våren 1999: Kommentert tabellrapport*, Statistics Norway, Oslo-Kongsvinger.

Statistical Yearbook 1994, Statistics Norway, Oslo-Kongsvinger.

St.melding 31: Om distrikts-og regionalpolitikken (1996-97) Kommunal og arbeidsdepartementet, Oslo [available at http://www.odin.dep.no].

Storstad, O. and Haugen, M.S. (1997) *Nøkkelinformasjon om kvinner i landbruket*, Centre for Rural Research, Trondheim.

Søbye, E. (1993) *Offentlig omsorg. Sosialt Utsyn 1993*, Statistics Norway, Oslo-Kongsvinger.

Chapter 5

Feminization Trends in Agriculture: Theoretical Remarks and Empirical Findings from Germany

Heide Inhetveen and Mathilde Schmitt

'Women are an advantage': The (Re)discovery of Female Farmers by Politics

Politics has re-discovered women farmers. Whereas agriculture at the end of the 1960s was considered to be a man's domain par excellence, the position of women in agriculture has progressed for at least a decade to be seen as the promising factor in agriculture and the motor of rural development. This re-evaluation of the role of women was put in writing and ratified by many nations at the 1992 Conference for Environment and Development in Rio de Janeiro, where it was stated in the so-called Rio Declaration on Environment and Development. Article 20 of this declaration points to the vital role women play in the management and development of the environment and calls for the broad participation of women in sustainable development.[1]

Rural women and female farmers have also been given more attention in political discourse within the European Union. 'The future of the rural regions depends to a large extent on the role women are delegated', according to a report written for the European Commission in 1994 (Flesch, 1994). In Germany the contribution of female farmers to the income of the families and to sustainable improvement in the quality of life in rural regions was explicitly mentioned in the Nutritional and Agricultural Policy Report (Ernährungs- und agrarpolitischer Bericht der Bundesregierung, 2003, p.65), which is published annually by the German federal government. In this report, it is stated that women secure the existence of their farms and strengthen employment and the economy of rural

---

[1] http://www.unep.org/Documents/Default.asp?DocumentID=78&ArticleID=1163 (download 21.01.2004).

regions, with rural women regarded as the 'trailblazers' in dialogue between producers and consumers. Added to which, female farmers appear to represent multi-functionality in agriculture in a special and specific way.[2]

The Feminization of Agriculture: The Career of a Rural Sociological Concept

Already by the beginning of the 1970s, the changed role of women in agriculture had been registered in agricultural sociology, and a new term was developed to describe the phenomenon; namely, the feminization of agriculture. Corrado Barberis was the first agricultural sociologist to transfer this term from general to rural sociology, and used it to describe developments in the agrarian sector. Yet women always played an important role in European agriculture, in particular in Germany: 'None of the countries in the EEC has shown greater skill than Germany in establishing women in the position of pre-eminent importance in agriculture' (Barberis, 1972, p.15). Even so, for Barberis the term feminization specifically describes 'the increase in absolute terms of the number of women working in agriculture', on the one hand, and 'more importantly the slower exodus [of women] from farming', on the other (Barberis, 1972, p.10). Hence, the feminization of agriculture can take place anywhere, even if to a different extent by country and region. 'In general, however, feminization reflects a sign of weakness' (Barberis, 1972, p.10). This is because economic crises are seen to have led to an exodus of 'the most courageous and the toughest' to the cities, leaving behind a residual population that did not remain in agriculture out of conviction but rather due to a lack of alternatives. In this sense, agriculture can be conceived as turning into 'the province of the physically, or at least economically, weak' (Barberis, 1972, p.10), primarily women and elderly men. In this process, Barberis sees the feminization of agriculture taking three forms:

- substitution (taking over activities because economic development allows men to disdain them)
- integration (when women do work ostensibly considered traditional for their sex)
- competition (when women vie with men for equal employment opportunities and in all aspects of social and political life)

With respect to the 1970s, Barberis discovered all the above substitutive forms of feminization as characteristic of farms that went through the transition from full-time to part-time farm status. He found this form of feminization especially marked in Germany, where women worked in agriculture in inverse proportion to the size of farms. He held that 'the lack of training, particularly occupational training among women' was primarily responsible for the substitutive form of feminization (Barberis, 1972, p.28). Due to this low level of vocational training, the role of women was seen to be limited to 'small undertakings', with associated consequences of low earnings, the exploitation of women's strength, illness and

---

[2] Information No. 28/2 of 9 July 2001, http://www.verbraucherministerium.de/pressedienst/pd2001-28.htm.

premature old age. According to Barberis, 'competition' could only be observed to a slight degree.

At the end of the 1970s, Cernea (1978) used the term 'feminization' to describe a phenomenon in socialistic agriculture, namely an unintended and not prognosticated consequence of the centrally planned industrialization and collectivization of Romania: 'The analysis of the occupational structure of Romania's total and agricultural labour force reveals immediately the occurrence and extent of a major unpredicted phenomenon: *the large increase in the number of women working in agriculture.* An unusual concept has been coined to describe this process: the *feminization* of agriculture' (Cernea, 1978, p.107f). Whereas Barberis focused primarily on feminization as a result of crises in agriculture, Cernea (and others[3]) regarded it more generally, as a consequence of industrialization and intensification (i.e. of modernization processes). Both authors stressed the fact that the feminization of agriculture does not necessarily lead to an improved situation for women. Rather women's role plurality can lead to demands on their time and efforts that are too taxing. Even if this feminization proves to be a concomitant phenomenon to positive economic progress, the 'traditional submission to their menfolk' continues. As Cernea prognosticated: 'It appears that a long time will be needed to alter both their economic roles and situation in the peasant household and the cultural values that are hindering change' (Cernea, 1978, S.120).

Barberis and Cernea indicate a correlation between the feminization phenomenon and the structural transformation in agriculture that had accelerated in the 1970s. A further insight on this concept came from the work of Inhetveen and Blasche (1983), who analyzed in the effects of modernization on (small-holding) agriculture in Germany. They used the term the 'feminization of agriculture' to describe expansion in the range of farm tasks for which women were responsible when their husbands worked off the farm. A positive effect of this development was the increased assumption of farm decisions by the women. Yet the exploitation of female farmers resulting from this new gender-specific division of labour was portrayed as negative (Inhetveen, 1982).[4] Pfeffer (1989) evaluated this thesis by analyzing nationwide data for the Federal Republic of Germany. The results 'do support the argument that a feminization of production takes place on part-time farms'. But women did not assume new fields of responsibility, 'instead, it [feminization] is the result of women assuming responsibility for a larger share of those tasks for which they already had shared responsibility' (Pfeffer, 1989, p.60). The question that arises is whether the 'integrative feminization' Pfeffer identified excludes substitutive aspects, as registered by Inhetveen and Blasche. One may

---

[3] Such as Abdelali-Martini *et al.* (2003) and Song and Jiggins (2003).

[4] Van Deenen (1983, p.1.26) also used the term 'feminization of agriculture' in connection with an intercultural comparison of women working in agriculture in Austria, France, Germany, Hungary, Poland and Sweden. He found the phenomenon primarily present on very small holdings at the limits of subsistence and on part-time farms. As a result of the feminization of agriculture, the social recognition of women rose, along with their work burden.

also suggest that the various forms of feminization, as differentiated by Barberis, can take place and be weighed and interlinked differently according to temporal and local conditions.

All in all, agricultural sociological research in the 1970s and 1980s agreed that the feminization of agriculture in 'socialist' and capitalist countries did not lead automatically to a more socially and politically equal status for women. But it did lead to a greater work burden. Research on developing countries came to a similar conclusion but accentuates the responsibility of agricultural policies, institutions and programmes (e.g. Safilios-Rothschild, 1991).

Since the 1990s, when structural change in agriculture accelerated to become a structural revolution, new occasions have arisen for rural gender research to deal with the concept of the 'feminization of agriculture'. At first glance, the 'competition' between genders that Barberis described as the hardly existent third form of the concept seemed to be taking shape. Thus, Ventura (1994, p.27) spoke about the development in Italy of a 'progressive feminization'. In Italy, the exodus of men from agriculture and the growth in part-time farming was linked to an increase in the number of women-owned farms and to a 'growing professionalism' amongst women. That, however, did not necessarily signify that women competed directly with men. Some of them would go 'beyond imitating the male pattern or the traditional pattern in which women either depended on men or replaced them'. In particular, young women with a higher level of education who were farm managers were 'motivated to work in agriculture by a longing for new lifestyles and habits that they actually can achieve in self-employed agricultural activities' (Ventura, 1994, p.33). According to Ventura, such women should no longer be referred to as farm women or women farmers but rather as female entrepreneurs. They go their own way in the process of a 'feminization of agriculture'. They reorganize available resources in an innovative way and try to convert household activities into market activities. Ventura postulated that these women farm managers 'might well become the spark that will generate endogenous development processes' (Ventura, 1994, p.38). Similarly, Bock (1994) discovered that women on Umbrian farms contributed in non-established ways to so-called 'non-agricultural activities', such as agro-tourism, direct-selling and processing, so offering survival during agricultural crisis.

Whereas the growing professionalism of Italian female farmers tended to be directed toward 'non-agricultural activities' that took place on the farm, in Norway women's activities tended to lead them outside agriculture. Thus, Norwegian researchers have found a 'masculinization of agriculture' within their country (Almås and Haugen, 1991; Blekesaune et al., 1993). This is associated with the development of a tertiary-sector based society, which improved job opportunities for women outside agriculture and facilitated their leaving the farm. That shifted gender-specific relations between the labour force to the favour of male manpower on farms. On the other hand, this 'masculinization of agriculture' contributed to a feminization of the gainfully-employed population within the general labour market; which is one of the most outstanding developments of this century across the whole of Europe, according to the sociologist and labour-market researcher Margaret Maruani (1997, p.49).

Feminization: A Re-institutionalization of Gender Differences at a New Level

Margaret Maruani (1997, p.50) established, with respect to labour markets in the European Union, a decrease in the predominance of men: 'The male hegemony in the world of work... no longer exists'. The number of gainfully employed women rose not only during economic booms but also during economic crises (e.g. in the 1970s), which is an indication that initiative was taken by women themselves. Maruani emphasized, on the other hand, that the feminization of the labour market did not signify that a really mixed working world had evolved. The number of women working in 'female occupations' did indeed increase, but 'male occupations' remained largely inaccessible to women. Maruani sees an important reason for this lying in the fact that the current definition of qualifications cements the gender-specific hierarchy. This is because, at each stage of the modernization process, only a small difference between female and male tasks was needed to maintain the gap between a qualified 'male occupation' and an unqualified 'female occupation'.

The sociologists Bettina Heintz and Eva Nadai (1998) also dealt with the development of the labour market. They argued that the increasing inclusion of women in the labour market led to a de-institutionalization of gender relations. This, however, did not signify that traditional differences between the genders had become irrelevant but rather that they are more and more dependent on specific contextual conditions. The reproduction of gender differences has shifted, so gender differentiation today is increasingly actively established and symbolically reinforced or, as the case may be, established by means of indirect and at first glance gender neutral regulations (Heintz and Nadai, 1998, p.88). In this regard, these sociologists speak of a 'contextual contingency' of gender differences. This de-institutionalization hypothesis for gender differences and their reproduction through action and symbolic levels needs qualitative verification in many areas of society.

Issues, Methods, Data

The above short review of the concept of feminization in rural sociology, as well as, in certain contexts, in gender research, brings up numerous questions that in our opinion are partially the result of imprecise terminology.

First of all, all analysts agree that a radical change in gender relations in agriculture can be observed, which is primarily based on women's changed behaviour, with change linked to structural change in agriculture; and so to the accompanying crisis phenomena. While many researchers use the term 'feminization of agriculture' to describe remarkable adjustments in gender relations, for others the term the 'masculinization of agriculture' is more appropriate. But, does this really signify counter-development? Does this not mean that one case refers to the initiative of women and the other to the consequences of their activities? Does feminization describe a quantitatively tangible shift in the gender labour market (and/or property relations) in agriculture in favour of an increase in the absolute number or a shift in the relative percentage of women in the workforce? Or does feminization signify qualitative change in the gender division of labour and its consequences? More clarity with respect to terminology could avoid these and other questions from the beginning.

However, there is a level in the feminization discourse that has been called for but has been hardly addressed. This is the level of individual and family-farm actions in the context of feminization.[5] In this context, the following questions pose themselves: What are the correlations between feminization processes and family/farm biographies, especially in the transition from full-time to part-time farming? How do feminization processes correlate with the 'curve of destiny of the farm family' (Planck and Ziche, 1979, S.300f)? How are differences between substitutive, integrative and competitive forms of feminization that are made in the literature on the subject in the everyday activities of the families and their male and female members presented? How can one concretely conceive the transformation of reproductive functions into commodities? Under what conditions do feminization processes take place? Do the economic and social roles of women in agriculture change in the process to form new orientations for their activities and new lifestyles? Does a so-called 'progressive feminization' (Ventura, 1994) lead to a 'feministic consciousness' (Lerner, 1993)? Do new personal arrangements or new social networks evolve? Do feminization processes also contribute to a new (family or public) esteem for the work carried out by women? Does overburdening and self-exploitation accompany 'progressive feminization' and constitute too high a price for women farmers? Will institutions of gender differences in agriculture be so weakened that a de-institutionalization of gender differences occurs? Does feminization reflect a weakness of agriculture, so it represents a survival strategy in the crisis-prone development of farms, or is it a contribution to the innovative reorientation of farming?

In order to answer these questions, the micro-level actions of agricultural males and females have to be taken into account and analyzed. To achieve this we draw upon data from two large-scale studies. These studies were carried out in 1977 and 1997. They involved analyzed changes in the orientation and patterns of action of female farmers in Bavaria, a region in Germany (i.e. a country in which, according to Barberis (1972), women always played a special role). The research design for the 1997 follow-up study provided the opportunity to correlate the socio-economic

---

[5]   Ventura (1994, p.38) in particular points to this gap: 'This seems a good reason to observe in more detail the behaviour and the choices made by these new women entrepreneurs as well as the new relationships arising within the farm families,...'. We interpret Barberis (1972, p.11) in similarly warning against an abstract treatment of the feminization of agriculture: It '...is not a neutral phenomenon, or one which can be considered in the abstract. It corresponds to a profound logic in the overall patterns of employment ...'. Abdelali-Martini and colleagues (2003, p.92) also refer to the changeability of gender roles 'in response to the national and even global contest' and argue that: 'These subtle changes can only be shown by appropriately detailed disaggregated studies'.

situations, identity constructions and agencies available to women in our target group with their experiences during the period of agricultural restructuring that occurred in the last two decades of the twentieth century. Using a semi-standardized questionnaire in the year 1977, 134 farm women between the ages of 18 and 66 were interviewed and questioned about various aspects of their biographies, their everyday lives, their work and their expectations with respect to the future of their farms (Inhetveen, 1982; Inhetveen and Blasche, 1983; Inhetveen, 1990). In 1997 we visited 128 women – only six women had died since the first study – and collected findings on changes and their effects, once again using a semi-standardized questionnaire. Biographical interviews were used to analyze how social and structural changes during the past 20 years were reflected in the individual lives of farm women. The theoretical frame used in the study was based on the theory of creativity of action developed by Hans Joas (1996). The project presents the possibility of answering the questions posed above in detail, with a view to exploring links between individual and family biographies as well as regional and social change.

The Feminization of Agriculture: The Case of Martha B

Farms in the Hands of Women: An Exception and a Challenge

When we interviewed Martha B in 1977, she was 30 years old, married and had five children. Her eldest daughter was 11 and her youngest son two years of age. Together with her mother, who was 62 at the time and still worked actively in the house and on the farm, she had to provide for eight people, four males and four females. Martha B's farm was situated in a highland region and belonged to the middle-sized farm category with its cultivable area of 13 hectares. The cropping on the farm was very diversified. The farm still had 28 cows but had given up keeping pigs except for two fattening pigs. The amount of machinery on the farm appeared to us to be rather large in comparison with other similarly structured farms. Two reasons were possible. On the one hand, Mrs B's husband had an administrative position in a local machinery pool and, as a result, was well-informed and up to date about the technical possibilities for rationalizing a farm. On the other hand, he had married into the farm. Farmers who had not inherited their farms felt – according to our observations – more pressure to attain status than their counterparts. By being particularly open to innovations in their activities, they attempted to compensate for their 'biographical blemish'.

A traditional institution in Germany in the 1970s was still male succession on the farm; a tradition that had been strengthened by the national socialistic hereditary farm policy. A certain 'de-institutionalization' in the form of female succession on the farm was a necessary, unwelcome exception that people tried to avoid. In the case of the female farmers we interviewed, however, the percentage of women who had inherited farms was astonishingly high, at just under 25 per cent, which was a consequence of the lost lives of their brothers during the Second World War, who would otherwise have inherited the farms.

The feminization of succession on the farm did not only have considerable consequences for husbands who had married into the farms. In addition, patrilineal succession of the family name led to the danger that the generations that had owned, built up and worked on the farm would no longer be remembered. However, the practice of establishing a 'house name' in many regions in Germany, which passes from generation to generation independently of the name of the current occupants, avoids the 'danger' caused by female farm succession.

Yet for women who inherited farms, the consequences were considerable. Martha B was raised to be the heir of the farm as soon as it became obvious that a male heir was not to be expected. She was included in all the tasks of the farm, which limited her 'leisure' more than her younger sister. She had no time for girlfriends. After completing a vocational school, she had to work full-time on the farm in order to replace her father who had died young. She married comparatively young, to a man who had also grown up on a farm.

Her education to 'think in terms of the farm' was very successful in the case of Martha B (as in the majority of cases for female farm heirs). Martha B identified herself strongly with the farm, with her occupation and with her role as a female farmer. She was especially proud to be her 'own master' – the same as her husband. The additional income earned by her husband, who worked part-time in the local machinery pool, was considered a temporary expedient. The married couple's great hope was to return the farm to its former status as a modern, full-time dairy unit. 'At the time, we thought that we would be a full-time farm in a year once again, that was [how] things looked'. In order to achieve this goal as soon as possible, they opened up a distillery as an additional 'para-agricultural' income source.

Martha B considered housework an independent occupation that she wanted to do professionally. Together with her mother, she kept a large garden, preserved fruits and vegetables, and sold products from the farm and garden, such as milk and fruit, to customers who bought directly from the farm. Her pluriactivities frequently brought her to the limits of her physical endurance. She had hardly any time left for herself and her hobbies, handicrafts and crochet work. The fact that she assumed a leading function as the 'Ortsbäuerin' (local female farmer representative) in the local farmers' association, and arranged and participated in the association's events, was an indication of another important, traditional rural farm institution; namely, the family or partnership production (i.e. cooperation within the family along the lines of a traditional division of labour between the farmer and his wife, and between older and younger generations). As in the case of most of the female farmers we interviewed, Mrs B carried out a twofold-socialization programme when raising her five children. On the one hand, the children should learn 'to think in terms of the farm'. The purpose of this orientation in their thinking and activities was to make them feel obligated towards the farm for all of their lives. This represents another institutionalized practice in rural agriculture. Thus Martha B attached great importance to the fact that not only the potential farm heir but all her children felt bound to the farm and family and were familiar with tasks and work on the farm. On the other hand, she encouraged the children to learn about other occupations 'in accordance with their individual capabilities', in view of the uncertain future of farming.

Intermezzo 1977-1997: The Agricultural Scene Shift, Also for Women

As a result of the reorganization of the milk market, German reunification and GATT negotiations, the farm sector underwent a structural shift whose dynamism and drama has no parallel in the history of agriculture. Economic survival became precarious, even for farms that looked hopefully toward the future in the 1970s. Living and working conditions for farm women changed drastically; with their continued existence in agriculture more strongly challenged than ever before.

The total number of farms in West Germany between 1977 and 1997 declined from 858 700 to 470 300. This means that 45 per cent of farms were given up. The number of family farm workers sank from 2 287 100 in 1977 to 1 210 100 in 1997; a decline of 47 per cent (Statistisches Bundesamt, 1979, 1998). In our sample, only 29 out of 134 farms (22 per cent) had given up operations completely. In addition to these farms, three development paths were identified. These broke down into 28 innovative or expanding farms (21 per cent) that faced the competition in agriculture, 39 farms (29 per cent) that strived to keep what there is and 38 farms (28 per cent) that dropped some of their activities. The frequently quoted argument that it is necessary to 'get bigger or step aside' was only partially confirmed.

The decline in the number of farms was accompanied by an increase in importance of family farms being run part-time. In West Germany, the percentage of such farms rose from 39.3 per cent in 1977 to 47.8 per cent in 1997; in our sample, from 43.3 per cent to 69.2 per cent. The transition to part-time farms was not only a strategy on farms that were reducing their activities, but could be found in all development paths. Only in a few cases could the shift be attributed to new gainful employment on the part of the woman farmer and/or her husband. More often than not it occurred when the farm was handed over to the next generation.

Male Farm Succession with Expansion and Commercialization of the Female Economy

Martha B was one of 128 female farmers we revisited and interviewed in 1997 after a span of 20 years. All in all, Martha B's twofold strategy in the socialization of her children had proven successful. Her daughters worked in the field of tourism or as a nurse. The heir of the farm had successfully completed vocational training to become a 'Landwirtschaftstechniker' (agricultural engineer), was married and had three small sons. A second son had become a carpenter, and the youngest son had just passed his higher secondary-school examinations ('Abitur'). Despite their vocational orientation outside the agricultural sector, all of her siblings maintained their bonds to the farm; a fact that made Martha B especially proud. The children did not 'perch together on a hen-roost' Martha B assured us. Each went her/his own way and led her/his own life. On the other hand, the children were 'at all times' available and willing to help on the farm whenever they were needed, even on weekends.

If we consider the fact that Martha B and her husband worked full-time on the farm and her mother was still quite active until recently, then we can say that the family and partnership production was strategically maintained. This pattern was also pursued by Martha B's daughter-in-law. She gave up her job as a home economics teacher to work on the farm and care for her three children along with her mother-in-law. The uniting of the personnel resources of the family was necessary if the farm was to remain as a competitive production unit. That this strategy succeeded is reflected in the fact that this farm was one of the 28 farms in our sample that had noticeably expanded their production since 1977. The cropped area had been increased by renting and buying land, so 40 hectares was now available to the family (i.e. more than three times the original area). Nevertheless, the, at the time realistic, expectation in 1977 that it would be possible to convert the farm from a part-time to a full-time unit had been destroyed by the events and setbacks of the 1980s. The main goal of increasing the milk quota and expanding dairy activities could not be achieved in view of the new agrarian-policy framework: 'It was the worst disappointment in my life'. But the family did not give up. It looked for new perspectives instead. The result was, in Martha B's words, 'still a farm, but it is structured completely differently now'; a statement that could be interpreted to mean a small-scale structural transition or a break with the past. In fact, the new arrangement was based on a dual strategy.

The first element of this strategy was that, instead of pursuing a transition to a full-time farm, expansion of off-farm activities occurred. Thus, the heir to the farm took advantage of the possibility to improve his education and get a job in the accounting office at the local farmers' association. Alongside this work, he took care of farm work or the supervision of construction work on the farm in the evening. The close relationship between farming and his work in the accounting office provided many synergic effects in the form of information on the latest agrarian policies, knowledge about the experience fellow farmers' survival strategies and the possibility of wining customers. The second element of the strategy was that, instead of increasing primary production, emphasis was placed on expansion and commercialization in the 'female economy'. In this strategy, Martha B and her daughter-in-law worked together, utilising available materials, personnel and qualifications in the house and on the farm to develop a new concept for the production and sales of their own farm's products. In addition to hard liquor, they produced different types of liqueur, made out of products raised on the farm, using a process they developed themselves. In addition, products that had formerly been primarily produced and processed by the women on the farm, within the framework of a subsistence economy, were now commercialized. The selling of these products was undertaken by the women, or sporadically by other female members of the family or relatives. Martha B summarized these new activities as 'direct sales'. The commercialization of commodities produced by the women demanded re-planning the future and, in accordance, new investment. The distillery had to be renovated, a room had to be created to process alcoholic products, a larger farm shop had to be built and a bar had to be furnished for groups of guests to sample the different types of alcoholic drinks. Martha B and her daughter-in-law considered baking bread and selling this directly along with beef. They were receiving customers on the farm not only from the near vicinity but also in the guise of large parties of tourists.

Agricultural Policies Continue to Change

Since 1997, the number of farms and farm workers has continued to decline. In West Germany, there were 416 672 farms in 2001, which employed a total of 1 162 000 workers. In the whole of Germany, there were 448 936 farms and 1 323 700 workers. Overall, the number of women employed in agriculture has declined. An unexpected change, however, can be seen amongst family workers, for the number of male workers has fallen more dramatically than female workers. The relative increase in the percentage of women, from 36 per cent to 44 per cent of full-time workers, and from 60 per cent to 65 per cent of family members working part-time in 2001 indicates a (quantifiable) trend toward feminization. Also rising slightly was the percentage of women farm owners from 8 per cent to 9 per cent. These developments are accompanied by a further increase in the number of part-time farms (Statistisches Bundesamt, 2003).

Epilogue

A year after our second interview, the farm was visited once again in another research context and Martha B and her daughter-in-law were interviewed for a third time.[6] In the meantime, a fourth grandson had been born, and Martha B's husband had passed away. Martha B had become an 'Altbäuerin' (senior female farmer) and lived in a four-generation arrangement on the farm, which had been passed onto the oldest son. The traditional patrilineal succession was again intact. Presumably it will continue into the next generation as the young farmer and his wife have four small sons.

In view of this changed family constellation, the decision to expand the 'female farm economy' seemed to be confirmed by the facts. The family planned to expand its successful 'direct-to-customer sales', although competition on the local markets had increased considerably. The dairy husbandry had been reduced to suckling cows and young stock was kept on pastures, while the former cropping activities had been almost completely replaced by the cultivation of flax, which demanded little labour. Martha B was noticing her physical and psychic limitations more – also as a result of her husband's demise – and warned her daughter-in-law she should take better care of herself. Leading from this, the two women deliberated over whether or not to limit their extensive subsistence-farming activities in order to find more time to dedicate to their successful market production.

---

[6] We would like to thank Juliane Waechter for giving us permission to use her interview.

Facets and Consequences of Feminization: Interpretation and Discussion

Feminization as a Farm Biography Necessity and Individual Option

In the case-study presented here, the feminization process can be observed on the basis of three striking factors in the farm biography.

It began at first with a quasi-fateful necessity, when Martha B's parents did not have a male child and a female heir was designated to take over the farm as successor. After the early death of Martha B's father, three generations of women (Martha B, her mother and grandmother) ran the farm. As a result of the break with the institution of patriarchal ownership on the farm, a gender-specific division of labour could not be practised in the traditional way. The women had to do (nearly) everything. An early marriage was an obvious option for the young heir of the farm (Martha B), who was continually overburdened. That gave her the possibility of delegating the traditionally male part of farm work to her husband. This situation changed slightly when the husband began to work outside agriculture. As the farm was not restructured, Martha B increased her workload on the farm. This was, according to Barberis (1972), a form of 'substitutive feminization'. This took place as a labour-organization necessity, caused by the transition to a part-time farm. It was not because the husband disdained tasks. At the same time, work within the family increased because Martha raised (together with her mother) five children. This represented a difficult, 'double-burdened' phase in Martha B's 'curve of destiny'.

A new stage in the feminization of the farm arose when the long-desired wish to transform the farm into a full-time operation, with the focus on dairy farming, the husband's favourite area, was thwarted by political developments. Various options were considered, including giving up the farm. The decision had to be made and borne not only by Martha B's husband, but also by the successor and heir to the farm and his wife, who lived on the farm with their small son. The decision in the end was to expand the 'female farm economy' and to establish a secondary business. Dairy farming, which was still in the hands of Martha B's husband, was to be maintained at the same level. The situation of the farm heir, who worked off farm for an agricultural organization, was hardly affected. The wife of the farm successor had the most important changes. She had qualified with a vocational education and had an interesting and well-paid job. Without her cooperation and her labour, the successful expansion of the secondary business was not imaginable. Why did she decide to give up her job?

As with Martha B, the daughter-in-law had been socialized in a rural farm tradition and had been raised to think of the farm first. It was important to her to maintain the farm, even if it had not been passed on and would – in accordance with the patrilineal succession – become her husband's farm. 'You are your own boss. You can decide yourself. You do it... for yourself... that is somehow an entirely different feeling'. As in the case of many young people raised in the farm tradition, she preferred to work on the farm: 'Somehow I prefer to work at home'. After considering the fact that she could accompany the progress and growth of her children better that way, she was more content and happy to invest her domestic economy qualifications in the farm. This strengthened her decision to take part in the 'female secondary business'.

In summary, the feminization of agriculture in the form of commercializing the 'female economy' has, within the biography of these people, families and farm, to be interpreted more as a last resort for the farm than as a freely developed decision. In other words, it reflects, on the one hand, the economic frailty of the farm and the agricultural sector, as Barberis (1972) noted. At the same time, it also reflects the strength of individual people and traditional family economy institutions. It represents – not only in this case study – a well-deliberated option that bundled farm and farm family resources cleverly in order to maintain the farm. With respect to this form of the feminization of agriculture, a de-institutionalization of the gender differences is not – as it would be in the societal labour market (Heintz and Nadai, 1998) – the important factor. To the contrary, the traditional institutions, in particular thinking in terms of the farm, farm socialization and family cooperation, remain to a decisive extent attainment oriented.

Manifestations of the Feminization of Agriculture

In the study carried out in 1997, a 'vacation on the farm' was an important income alternative (being offered on 20 farms, i.e. on 19 per cent of the 105 farms that still existed from the 1977 sample), as was the processing and sale of a farm's own products (which occurred on 80 per cent of farms). Other traditional women's jobs, such as nursing (e.g. taking care of elderly people or running a boarding-house for dogs) and housekeeping, were also very popular. Female farmers were thus able to take advantage of the knowledge they had gained through experience, as well as discover and take advantage of income sources in the tertiary sector – above all in regions in the vicinity of towns – through the professionalization of their activities (e.g. in opening a café on the far, or a party-service or a washing and ironing business).[7] For 'vacations on the farm' more and more holiday apartments were offered, instead of bed and breakfast, which meant less work for farmers' wives.

In addition to 'professionalization', 'cooperation' and 'centralization' were key words in development. Group-oriented diversion and recreation activities for the guests were developed. For such activities, advertising was assumed centrally by the regional tourist office. Committed female farmers promoted their regions all over Germany at fairs and markets.

Prerequisites for the Commercialization of the Female Economy

The development and expansion of a secondary business can take advantage of existing spatial and material resources, although as a rule it requires further investment and construction measures. At the same time, the necessary recruitment

---

[7] The Bavarian Agricultural Ministry reacted to this phenomenon by providing a special extension service: 'Hauswirtschaftliche Dienstleistungen profilieren sich auf den Märkten'. The effect was the establishment of a 'Hauswirtschaftliche Fachservice-Organization' in almost every county.

of other family members for such activities creates impulses to expand social resources. Thus, cooperation and networks ensue within the family and amongst relatives.

In our case study, the core of the activities was a secondary business consisting of a mother-in-law and daughter-in-law. In other families, the team consisted of a mother and daughter. In any case, these feminization strategies had a female core. Cooperation took place according a traditional model, as 'teamwork', as Martha B described her working relations with her daughter-in-law. The daughter-in-law attributed successful cooperation to complementary characters, similar perceptions of reality, clearly defined delimitations of tasks, good reciprocal perception and acceptance of each other's strengths and weaknesses. It appears as if this form of the feminization of agriculture can positively influence the traditionally rather complicated mother(in-law) and daughter(in-law) relationship (Schmitt, 1988; Silvasti, 1999; Wächter, 1999). As is traditionally the case between a farmer and his wife, the two women met each morning to discuss key tasks for the day and distribute work for the day and/or week.

The two women were helped by other female relatives when marketing their products. One of Martha B's daughters helped sell products at the market, and the mother of the daughter-in-law took care of the grandchildren. Martha B's second daughter (who worked in a tourist office) and her sister (who had a bus enterprise) helped them find interested customers. The commercialization of the female economy was based, in other words, on existing social resources, that led to a new network between female members of the family and their relatives. This was reciprocal. The feminization of agriculture, according to our hypothesis, remains embedded in the traditional family enterprise. It necessitates and promotes family enterprise thinking and behaviour.

Feminization also necessitates and promotes the utilization of qualifications and resources. Through her 'training on the job', Martha B gained comprehensive qualifications in both the domestic and the farm and economic sectors at an early age. And since women are also responsible for family 'well-being', as a rule they have qualifications and resources to provide services and leisure activities that meet customer requirements within a 'wellness society'. As Yvonne Verdier (1982, p.35ff) reminds us: 'Female farmers have to be able to do everything'. Martha B's daughter-in-law had learned the economy of women on her parents' farm and had qualified vocationally at the school for domestic economy – as many farmers' daughters do. The daughter-in-law caught up with her mother-in-law's special knowledge with respect to the production of hard liquor and liqueurs by working closely with her, by observing, asking to be shown how to do something and imitating. The two women attended seminars to perfect their spectrum of activities, from the cropping of raw materials to the choice and design of pretty glass bottles. Their work was characterized by a growing professionalism. Moreover, the two women complement each other in their joy over creatively designing their products and experimenting with new products and ideas.

The feminization of agriculture in the form of the commercialization of various fields of the 'female economy' increases the number of activities female farmers enjoy (Inhetveen and Blasche, 1983, p.209ff). The desire to organize and control their own work and to 'be their own boss' is for Martha B and her daughter-in-law more important than anything else. The development and expansion of the female economy offers these two women a framework in which they, like 'female entrepreneurs' (Ventura, 1994), can unfold their experiences, preferences and passions.

As self-employed people, they are less affected by the gender-specific inequality of the labour market than employees. But on the other hand, they are exposed to the fluctuations of the market in a new way, in their role as suppliers and sellers of their own products. They have continually to watch the market and look for new niches and follow the latest fashion. They need 'situational creativity' (Joas, 1996), which is both required and promoted by their activities: 'You just have to look and see what you can do or which direction to take, you have to be flexible... and have a feeling, a feeling for it. You just go in that direction'. 'Flexibility' is a term the two women repeatedly used to characterize their activities, whether referring to the process of one task alone or the organization of the farm as a whole. Flexibility was seen as a potential with respect to the development and expansion of the female economy and as something that let them look towards the future without qualms. One thing was clear to both women: 'A lot is going to change... until that is all ready: Well, I think something will always change anyway'.

The Feminization of Agriculture in the Public Eye

Martha B and her daughter-in-law received an unusual form of public recognition for their activities: 'The recognition for your work, when you notice that more and more customers come or that they are enthusiastic... they say, "That is good". And then they buy it... that's fun!'. Since they were ousted from local markets and the public sphere in the nineteenth and twentieth centuries, the only tasks and occupations for female farmers have been, and still are, tasks that earn very little money and very little public recognition. The sales of products that are processed by themselves brings women once again both money and recognition. The experience of having their own work recognized and valued both verbally as well as monetarily is an important attraction for women to conquer new branches of enterprise and to compensate for underestimation by their families and society of their achievements in agriculture (Fahning, 2001). The successful marketing of their products filled the women with pride with respect to their role as producers: '...and what I do with it, that is our idea and [it] is in our own hands. In our diligence, in our...'.

A correlation between the feminization of agriculture and a greater presence of women in the public domain has been described in the literature (Song and Jiggins, 2003). Our case study tends to corroborate this. At the time of her second interview, Martha B had been active in the local farmers' association for over 30 years, and had long been an active member in a church association and the local horticulture association. Added to which, she has just taken on a new function on the district executive committee of the rural women's association, for which she was organizing large rural women's meetings. But with her work in associations,

Martha B acted in (semi)public space, much as she did with her new secondary business, which depended on traditional customer contacts that are not established by money alone, but rather thrive on familiarity and trust. It is worth noting in this regard that her customer clientele and the membership network of organizations and associations overlapped. Good work in associations could thus have synergy effects for direct marketing, and vice versa. Her own work in associations could direct the flow of visitors. Hence, the increase in communal activities that Song and Jiggins (2003) observed as resulting from the feminization of agriculture in China could be interpreted and understood as well-directed network maintenance, which is performed with awareness of synergy effects.

Feminization and Gender Relations

In the 1970s, patriarchal power structures in rural families were still lived and accepted as natural practice. When Martha B's husband decided to tear down the baking house on the farm, which meant a lot to her as it was the former centre of women's work on the farm, she accepted this without protest: 'One must always give in a bit'. The discourse surrounding the 'campaign to achieve social equality for farm women', along with agricultural social reform and the introduction of so-called 'Bäuerinnenrente' (old-age pensions for farm women), has laid down an important legal and mental groundwork for consciousness raising amongst farm women, as well as for equal partnerships on farms.

With its commodification of the female economy, the feminization of agriculture could support women in their demands for a position with equal rights. It tends to equalize relationships (Giraud, 1999) and, as a result of public esteem, shifts power relations on the farm in favour of women. If secondary businesses become increasingly visible in the form of buildings, customer flows and increased revenue, while the significance of agricultural businesses on the farm diminishes, then men will be challenged to rethink their identities. Many are finding this difficult and insist on women's having a secondary role on the farm (Canoves and Villarino, 1999). More than a few farm women help maintain a patriarchal façade (Modelmog, 1994). Even evasiveness in interviews in response to questions touching on this issue can be interpreted in this light.

Yet only a few farm women see the chance for interchangeability of male and female roles, which is an opportunity, as Fonte and colleagues (1994) call it, connected with the commodification of reproductive functions. After they are no longer able to fall back on familiar institutions, such as the formal separation of the home economy and the market economy, male and female spaces, and the separation of public and private, women as well as men do not find it easy to interactively produce and symbolically affirm gender differences under the new conditions (Heintz and Nadai, 1998). Again, it appears that reorganizing the division of labour in agriculture is much easier than redistributing power between men and women (Janshen, 1989; Schmitt, 1997).

Feminization and Feminist Consciousness

In turning their home economy into commodifiable products, farm women like Martha B, not only have to develop creativity but also courage, because they are transgressing boundaries set by tradition and custom. Martha B, for instance, did not follow the custom of the region and planted different berry bushes for her liquor production, and she sat — together with her daughter-in-law — in one of her fields and plucked dandelion blossoms. For such acts, women in the village called her 'crazy' or laughed at her. Martha B did not let this bother her. On the one hand, she knows that change is necessary if you want to survive. On the other hand, this work allows her to enjoy nature while looking after her grandchildren.

Martha B is transgressing boundaries that are often set in rural society against innovation. New products and marketing from women provoke comments from villagers. Yet Martha B and her daughter-in-law have done much to integrate their actions into the rural family economy and rural society, so their 'crazy behaviour' will not have far-reaching consequences. The women take care that their emancipation occurs in harmony with their social environment. In another context we have described this self-awareness as 'relational self-concept' (Inhetveen and Schmitt, 2001). But our study did not identify any development of feminist consciousness, as defined by Gerda Lerner,[8] in the context of and as a result of the feminization of agriculture. Nonetheless, there are signs of self-confidence, as described by Ventura (1994), that successful women entrepreneurs tend to exude.

Self-Exploitation as the Price of Feminization?

The successful development and commodification of the female economy also have their price in an increased workload and extreme time pressure. The tendency toward 'self-exploitation', a characteristic of the rural family economy, is connected with the linking of entrepreneur and worker status. Farm women are especially vulnerable to this tendency because of moral values they have inherited. Relatively constant hard work, as with the male ideal of the industrial society (Rose, 1997, p.130), was a significant constituent of farm women's self-image and the image others had from them.

Lack of time limited Martha B's availability for both interviews. Like Martha B, her daughter-in-law does not count her working hours. Apart from difficulties in planning farm work in general, owing to unforeseeable events that can lead to a sudden increase in work, direct marketing and a customer-oriented time schedule add to time pressures. The ability to cope, according to Martha B's daughter-in-

---

[8] Gerda Lerner (1993, p.30f.) defined feminist consciousness as 'women's realization that they belong to a subordinate group; that they as a group suffer under a deplorable state of affairs; that their subordinate status is not preordained by nature, rather socially produced; that they must join together with other women in order to abolish this deplorable state of affairs; and finally, that they can and must work out an alternative vision for a social order in which women, like men, enjoy autonomy and self-determination'.

law, requires a specific 'Lebenseinstellung', or 'attitude toward life', that one develops only when one grows up farming.

If an illness, or even death, is added to the everyday workload, it can become particularly difficult and barely manageable. Women who work to their physical limits for a long time, or even beyond them, fall ill (Lasch, 1995; Wonneberger, 1995). In the case of Martha B, first a serious illness and then the sudden death of her husband gave her cause to stop and reflect. Now she regularly tries to 'take a break at noon for half an hour' and insists her daughter-in-law also takes a midday break.

Martha B expects further change in the future. This change should 'not be so extreme', not the big lottery win, nor a great mishap. With her current concepts, she feels prepared for such a development: 'After all, I'm flexible. I'm just happy every day, every week that things are going well, that everything is running smoothly. And we'll take it as it comes. One builds up, and one keeps building'.

Conclusion

This case study demonstrates that radical changes in agriculture and rural society can be successfully mastered by farm women to their satisfaction if they transgress boundaries that would hardly have been imaginable a few years ago, or which they would have refrained from out of fear of sanctions. These transgressions articulate both the notions of new individual patterns for organizing one's life as well as attempts to maintain the farm and the family enterprise in line with the development of the female economy. In the process, women (re)act not only as crisis managers. Their engagement should also be understood as a desire for self-realization and the realization of creative potential. As a negative consequence of the successful 'feminization of agriculture' one has to concede that a lot of the women must cope with an increased workload and extreme time pressure. Overall, the results of our study lead us to agree with Song and Jiggins (2003, p.287), who point out the positive side-effect of the feminization of agriculture, that '… the lessons learned offer promise that women farmers' experience, skills, and needs will be more respected as agricultural modernization proceeds'.

Our analysis is based on detailed observation and study of processes that lead to a commodification of the female economy. It refers to a case study whose particularity and generalizability can be demonstrated with a comparative analysis (forthcoming). In view of the heterogeneity of agriculture, including across regions in Germany and Europe, comparative studies need to be conducted to assess the feminization of agriculture in the biographies of other farm women.

References

Abdelali-Martini, Malika, Golden, Patricia, Jones, Gwyn E. and Baley, Elizabeth (2003) Towards a feminization of agricultural labour in northwest Syria, *Journal of Peasant Studies* 30(2), pp.71-94.

Almås, Reidar and Haugen, Marit S. (1991) Norwegian gender roles in transition: The masculinization hypothesis in the past and in the future, *Journal of Rural Studies* 7, pp.79-83.

Barberis, Corrado (1972) *The Changing Role of Women in European Agriculture*, Food and Agricultural Organization of the United Nations, Rome.

Blekesaune, Arild, Haney, Wava G. and Haugen, Marit S. (1993) On the question of the feminization of production on part-time farms: Evidence from Norway, *Rural Sociology* 58, pp.111-129.

Bock, Bettina (1994) Livelihood strategies: Women and the future of Umbrian family farms, in Margaret van der Burg and Marina Endeveld (eds.) *Women on the Family Farm: Gender Research, EC Policies and New Perspectives*, CERES, Wageningen, pp.83-90.

Canoves, Gemma and Villarino, Montserrat (1999) Rural tourism, gender and landscape conservation: the north of Spain and Portugal, in Gender Studies in Agriculture (ed.) *Gender and Rural Transformations in Europe: Past, Present and Future Prospects*, Wageningen University, Wageningen, pp.141-145.

Cernea, Michael (1978) Macrosocial change, feminization of agriculture and peasant women's threefold economic role, *Sociologia Ruralis*, 18, pp.107-124.

Deenen, Bernd van (1983) *Europäische Landfrauen im sozialen Wandel, Band II: Interkulturell vergleichende Forschungsergebnisse*, Forschungsgesellschaft für Agrarpolitik und Agrarsoziologie, Bonn.

Ernährungs- und agrarpolitischer Bericht der Bundesregierung (2003) hgg. *Bundesministerium für Verbraucherschutz, Ernährung und Landwirtschaft*, Medien- und Kommunikations GmnH, Berlin (http://www.verbraucherministerium.de).

Fahning, Ines (2001) *Frauen sind ein Gewinn: Beitrag der Frauen am landwirtschaftlichen Gesamteinkommen*, Niedersächsisches Ministerium für Ernährung, Landwirtschaft und Forsten, Hannover.

Flesch, Colette (1994) Vorwort in Braithwaite, Mary (1994) *Der wirtschaftliche Beitrag und die Situation der Frauen in ländlichen Gebieten*, Amt für amtliche Veröffentlichungen der Europäischen Gemeinschaften, Luxemburg.

Fonte, Maria, Minderhoud-Jones, Marilyn, Plas, Leendert van der and Ploeg, Jan Douwe van der (1994) The menial and the sublime, in Plas van der Leendert and Maria Fonte (eds.) *Rural Gender Studies in Europe*, Van Gorcum, Assen, pp.1-16.

Giraud, Christophe (1999) Understanding and misunderstanding between men and women in a farm tourism activity, in Gender Studies in Agriculture (ed.) *Gender and Rural Transformations in Europe: Past, Present and Future Prospects*, Wageningen University, Wageningen, pp.218-224.

Heintz, Bettina and Nadai, Eva (1998) Geschlecht und Kontext: De-Institutionalisierungsprozesse und geschlechtliche Differenzierung, *Zeitschrift für Soziologie*, 27(2), pp.75-93.

Inhetveen, Heide (1982) 'Nie fertig mit Anschaffen und Anpassen': Kleinbäuerinnen zwischen Tradition und Fortschritt, *Sociologia Ruralis*, 22, pp.240-263.

Inhetveen, Heide (1990) Biographical approaches to research on women farmers, *Sociologia Ruralis*, 30, pp.100-114.

Inhetveen, Heide and Blasche, Margret (1983) *Frauen in der kleinbäuerlichen Landwirtschaft: Wenn's Weiber gibt, kann's weitergehn...*, Westdeutscher Verlag, Opladen.

Inhetveen, Heide and Schmitt, Mathilde (2001) Two-thirds emancipated: Persistence and change in action patterns of farm women on German small holdings, in Agricultural Research Institute (ed.) *The New Challenge of Women's Role in Rural Europe*, Nicosia, pp.301-307.

Janshen, Doris (1989) 'So viel wert wie ein halber Bauernhof': Über die Feminisierung der Landarbeit und die Zukunftschancen für den ländlichen Raum, *Vorwärts*, 15, pp.26-31.

Joas, Hans (1996) *Die Kreativität des Handelns*, Suhrkamp Taschenbuch, Frankfurt am Main.

Lasch, Vera (1995) *Arbeit und Belastung bei Bäuerinnen*, Hrsg. von der Arbeitsgemeinschaft ländliche Entwicklung im Fachbereich Stadt- und Landschaftsplanung, Gesamthochschule Kassel.

Lerner, Gerda (1993) *Die Entstehung des feministischen Bewusstseins: Vom Mittelalter bis zur Ersten Frauenbewegung*, Campus, Frankfurt am Main.

Maruani, Margaret (1997) Die gewöhnliche Diskriminierung auf dem Arbeitsmarkt, in Irene Dölling and Beate Krais (eds.) *Ein alltägliches Spiel: Geschlechterkonstruktion in der sozialen Praxis*, Suhrkamp, Frankfurt am Main, pp.48-72.

Modelmog, Ilse (1994) *Versuchungen. Geschlechtszirkel und Gegenkultur*, Westdeutscher Verlag, Opladen.

Pfeffer, Max J. (1989) The feminization of production on part-time farms in the Federal Republic of Germany, *Rural Sociology*, 54, pp.60-73.

Planck, Ulrich and Ziche, Joachim (1979) *Land- und Agrarsoziologie: Eine Einführung in die Soziologie des ländlichen Siedlungsraumes und des Agrarbereichs*, Eugen Ulmer Verlag, Stuttgart.

Rose, Lotte (1997) Körperästhetik im Wandel: Versportung und Entmütterlichung des Körpers in den Weiblichkeitsidealen der Risikogesellschaft, in Irene Dölling and Beate Krais (eds.) *Ein alltägliches Spiel: Geschlechterkonstruktion in der sozialen Praxis*, Suhrkamp, Frankfurt am Main, pp.125-152.

Safilios-Rothschild, Constantina (1991) Women as a motor in agricultural development, in Gender Studies in Agriculture (ed.) *Gender Methodology in Agricultural Projects*, Wageningen University, Wageningen, pp.53-64.

Schmitt, Mathilde (1997) *Landwirtinnen: Chancen und Risiken von Frauen in einem traditionellen Männerberuf*, Leske+Budrich, Opladen.

Schmitt, Reinhold (1988) Hofnachfolger, weichende Erben und moderne Schwiegertöchter. Aspekte der internen Strukturveränderung bäuerlicher Milieus, *Zeitschrift für Agrargeschichte und Agrarsoziologie* 36(1), pp.98-115.

Silvasti, Tiina (1999) Farm women, woman farmers and daughters-in-law: Women's position in the traditional peasant script in Finland, in Gender Studies in Agriculture (ed.) *Gender and Rural Transformations in Europe: Past, Present and Future Prospects*, Wageningen University, Wageningen, pp.526-533.

Song, Yiching and Jiggins, Janice (2003) Women and maize breeding: The development of new seed systems in a marginal area of south-west China, in Patricia L. Howard (ed.) *Women and Plants: Gender Relations in Biodiversity Management and Conservation*, Zed, London, pp.273-288.

Statistisches Bundesamt (1979, 1998, 2003) *Statistisches Jahrbuch für Landwirtschaft*, Ernährung und Forsten, Wiesbaden.

Ventura, Flaminia (1994) Women in Italian agriculture: Changing roles and arising problems, in Margaret van der Burg and Marina Endeveld (eds.) *Women on Family Farms: Gender Research, EC Policies and New Perspectives*, CERES, Wageningen, pp.27-40.

Verdier, Yvonne (1982) *Drei Frauen. Das Leben auf dem Dorf*, Klett-Cotta, Stuttgart.

Wächter, Juliane (1999) Mothers, daughters and daughters-in-law as 'key-persons' on family farms, in Gender Studies in Agriculture (ed.) *Gender and Rural Transformations in Europe: Past, Present and Future Prospects*, Wageningen University, Wageningen, pp.549-555.

Wonneberger, Eva (1995) *Modernisierungsstreß in der Landwirtschaft oder was hat die abgepackte Milch mit den Bäuerinnen zu tun?*, Centaurus, Pfaffenweiler.

Chapter 6

Partnerships for Regional Development and the Question of Gender Equality

Christina Scholten

Introduction

In what sense is regional policy thought to ease everyday living conditions for women living in Sweden's rural areas? Despite the Swedish Government's recognition of feminist discourse and women's situations, regional development policy is characterized by favouring manufacturing industry and market competition. Despite feminist criticism, the national government, as well as regional and local governments, is . failing to adopt a true gender equality perspective in policy. In accordance with the dictates of globalization and neo-liberal values, regional development policy, as a means for equalizing economic distribution between regions, is 'abandoned' in favour of regions racing against one another for economic development. In this chapter I argue that the consequences of this policy are gender discriminatory. In particular, the idea of running a business is strongly gendered as a male activity (Sundin and Holmquist, 1989; Holmquist and Sundin, 2002), while women's interests in handicrafts and the service sector not are considered 'real' business activity, that are important to regional development and economic growth. A gender divided Swedish labour market and patriarchal structures are maintaining restrictions of great importance to women who want to be supported as businesswomen. In this chapter I present examples both from women who failed in their attempt to start their own business and from business advisors whose aims are to support women and men who want to start their own business.

Following Swedish membership of the European Union, new organizations for regional development emerged as an adaptation to international standards. These committees and their constitutions are of great importance when it comes to specifying what is considered valuable and needed for regional development and livelihood strategies according to gender. The constitutions of these partnerships are extremely important to what questions will actually be put on the development policy agenda. Yet there are only vague descriptions of who shall participate in these partnerships on regional development, other than that they shall involve cooperation between private and public actors. Yet, gender equality aspects should be acknowledged in accordance with gender equality legislation.

Regional development as an interest for politicians might be traced back to the 1930s at least. It was not, however, until 1965, that it became a field of politics for the Swedish government. The 1965 Act on regional development policy was a necessary attempt from the Swedish Government to respond to demographic situations in different parts of the country, as well as an attempt to transform and control development in general. Structural economic change, demand for more skilled labourers and the introduction of new technologies within industry, all helped speed up economic transformation in the years that followed (Berger, 1995). Also accompanying these changes, for decades out-migration had seen losses from less populated areas with weakening labour markets. Most notably, people moved to the industrialized areas in the south of the nation. Regional policy became a tool to 'control' this development and force industry to relocate to less populated areas. This policy had the aim of spreading wealth over the nation. When Sweden enlarged its welfare state and expanded the public sector, new labour markets were created all over the country. The public sector became and still is a labour market that is dominated by women. The critical point from the expansion of the public sector is that women had possibilities to obtain paid-work, even in remote areas, while children were provided with schools and day-care centres, and homes for the elderly were built. Living conditions and living standards became comparable across the country. Hence, regional development policy became a tool in the creation of the welfare state.

Compared with this positive picture, today the welfare state is being put under tremendous stress. There are constant debates on financing services provided by the public sector (Cameron and Gonäs, 1999). Enhancing the intensity of these debates, in the mid-1990s, Sweden, like most western economies, faced economic recession. Both men and women became unemployed in large numbers. The workforce in the public sector was reduced by 500 000. Women held many of these jobs. Consequently, in municipalities where the public sector dominated the labour market, reductions had severe implications for the municipality and, of course, for the ability of families to support themselves. The political leadership argued for several solutions. One of these solutions was that people should be supported in their decisions to start new businesses on their own. To many, both men and women, unemployment certainly was the prompt that was needed to start a business they had longed for. To others, the demand to start a business became their last trip from the labour market. In the penultimate section of the chapter I give two concrete examples of smashed business projects.

Regional Development Policy and Gender

Regional disparities and uneven development are characteristics all countries have to deal with. These are not new phenomena (Massey, 1994). In 1965 when regional development was founded as an object of national government policy (Berger, 1995; SOU, 2000, Prop. 2001/2002), and ever since that time, the issues of regional development have been largely about large-scale structures. These structures have been about migration, which led to depopulation in some parts of the country, and to overpopulated cities in others. Another structural aspect has been the location of industry and how to use governmental funds to re-locate private industries, state-owned industry and government agencies to less populated areas.

But regional policy has not been static, for its form has changed over the years. In the first two decades after it emerged as a political field, regional policy was primarily about distribution, in order to create somewhat equal conditions in governmental and municipal services throughout the country. By the beginning of the 1990s a new regional development policy had been outlined. For example, in the regional development proposal of 1993/94, competitiveness between Swedish regions was mentioned as a way to strengthen the economic position of the country and the welfare-base of Swedish society (*Bygder och Regioner i Utveckling*, 1993/94, p.140). This new policy reflected what were seen as a necessary adjustment to the European Union and to the forthcoming referendum on Swedish membership. In 1994 Sweden voted for membership of the Union and in 1995 became a Member State. By this time Sweden, like many other western economies, was on its way toward economic depression.

Despite the negative attitude toward membership of the European Union amongst those who lived in rural areas, and especially amongst women, membership gave opportunities to draw on the EU's Structural Funds. From this, Sweden managed to enlarge the regions of the country that were designated as being in need of structural change, with today's emphasis on inter-regional competition probably softened by funding opportunities afforded by Objective 5b monies for structural change and Objective 3 financing for social restructuring. One of the new rural regions that were identified as being in need of structural change was the Objective 5b area of Southeast of Sweden.[1] This region consists of 20 municipalities in the four counties of Östergötland, Jönköping, Kalmar and Kronoberg.

Objective 5b Southeast Sweden

The region chosen for this research is situated in the southeast of Sweden. The region was formed after Swedish membership of the European Union. There might be several dilemmas about creating a new region out of four existing administrative entities. One, of course, is that the municipalities that constitute the new region belong to different counties. Another consideration is the difficulty of creating a unified mission for local administrations that have not worked together before; indeed, in some cases, that have not had the opportunity to work together because

---

[1]   This area incorporates the municipalities of Boxholm, Kinda, Ydre, Valdemarsvik and Ödeshög in the county of Östergötland, Aneby, Eksjö, Nässjö, Sävsjö, Tranås and Vetlanda in the county of Jönköping, Borgholm, Hultsfred, Högsby, Mörbylånga, Vimmerby, and Västervik in the county of Kalmar, and Lessebo, Uppvidinge and Tingsryd in the county of Kronoberg.

municipalities are situated in different counties. A third issue that arises in this case is the implication of being identified as a region in need, for, in Sweden, regions in need of economic subsidies have principally been located in the north. With the new political agenda surrounding Objective 5b designation, politicians and administrators in Southeast Sweden had to forge a new identity and foster strategies amongst local and regional actors on how to benefit most from being designated as a region in need of structural change.

Figure 6.1 The Objective 5b Area for Southeast Sweden

The Southeast Sweden region was chosen in part because it was situated in the south, but also because a SWOT-analysis[2] revealed that people in this region were less educated, that manufacturing industry tended to be more traditional, that the demographic structure was skewed towards the elderly, that there was a lower Gross Regional Product (GRP) than the average, and so on (County Administrative Board of Jönköping, 1996). To capture key elements in the local labour market, its attributes can be divided into a 'private' male labour market of employment in manufacturing and engineering industries and a public sector that is dominated by female workers. Within Southeast Sweden, women are better educated than men but suffer more from part-time unemployment and part-time sickness leave. The gender system in this region was also brought into highlight through a survey on

---

[2]	SWOT is short for Strengths, Weaknesses, Opportunities and Threats.

regional policy and regional development projects, which found that women did not participate in nor initiate development projects. Since livelihood strategies and sustainability are basic needs for everyone, this research focused on women's ability to draw on subsidies in order to become self-sufficient and take control of their lives.

Structural Change

During the 1990s several changes in policy and society took place with Sweden. First, Sweden had a referendum on membership and then joined the European Union. Second, Sweden, like most western economies, was severely injured by economic recession. Third, strong arguments for the creation of more new, small companies were put forward by politicians. In a sense, all these changes were connected and impacted on people's lives in different ways and at different societal levels.

When Sweden applied for membership in the European Union, structural regional development funds helped speed up the reorientation of policy towards a regional development policy that encouraged increased competition between Swedish regions. EU membership negotiations also led to new regions being categorized as temporarily in need of regional development aid. In order to be able to secure Structural Funds from the European Union, regions had to complete a SWOT-analysis. This analysis was designed to identify the strengths and weaknesses of regions and signify in what ways regional development funding would improve development. A regional partnership, which involved a collection of both private and public sector actors, was supposed to decide on future development directions. Suddenly, politicians became more like the policy background. By contrast, salaried employees became more important, especially if they worked at the regional government level, as they were responsible for making invitations to partnership involvement and were important agents in discussions on the future of regional development.

At the beginning of the 1990s, the Swedish economy began to contract. Different programmes were organized in order to take care of people who were losing their jobs. These included adult education programmes, the expansion of the universities and higher education, special training programmes and extended efforts to make people start their own businesses. For the last of these, there were special campaigns aimed at encouraging women to start their own business. A reason why women, immigrants and young people were identified as potential businesswomen and businessmen was their lack of representation within existing 'business clusters'. But research had already shown that running a business is strongly gender coded as a male activity (Sundin and Holmquist, 1989; Holmquist and Sundin, 2002). One effect of this was that special businesswomen advisors were financed and organized by the Swedish Government in areas that were demarcated as being in need of structural change (NUTEK, 1996). Promoting women to start their own businesses was considered by the Government as one step toward stabilizing the economy as well as creating livelihood opportunities in rural areas, where the labour market traditionally has been weak (*Bygder och*

Regioner i Utveckling, 1993/94, p.140). My concern as a researcher of this policy
was to ask whether it is possible to get women to become businesswomen just
because there is a political ambition to establish more companies? This research
was also prompted by a desire to establish if women living in rural areas have what
it takes to create a business from which they are able to make a living.

Outline for a Feminist Theoretical Understanding

Sweden has come a long way compared with many western countries in terms of
formal gender equality. By formal equality I refer to juridical legislation. In
Sweden the welfare state does not tolerate discrimination according to gender;
same-sex couples have the same rights and responsibilities as heterosexual couples
living together. Prostitution is criminalized and the Government has made a
contribution toward putting the situation of battered women on the political
agenda. Still, everyday practices and the everyday lives of women and men in
Sweden indicate that we have to work on gender equality if we are to take it any
further. Women living in heterosexual relationships still have the major
responsibility for the home and the family, while women earn less than men for
comparable paid-work, and women exercise less power, both economically and
politically. Single mothers are also still dependent upon the welfare state. All in all,
this pattern can only be described as the persistence of patriarchal structures
(Walby, 1990), yet real efforts have been made to fulfil the intentions in the 1980
Act on the equality of women and men.

In my research I have used the concept of gender-system and gender-contract to
analyze how women are defeated in their attempts to provide for their own
livelihood strategies after they became unemployed. The concept of gender-system
is based on women's and men's different positions within society and, within that,
of women's general subordination. Here, culture, as seen in norms and traditions
within behaviour and values, are seen to impact on what is considered male and
female activities, respectively. Gender has its imprint on the whole of society,
although, to some extent, what this means differs between regions and countries as
traditions and norms vary.

In a Swedish context, Hirdman (1989, 1990, 2001) has argued that gender
conflict has been effectively suppressed by the Government, which has sought to
meet demands from women. The argument here is that issues that have been raised
by women's fractions within political parties or by professional women have been
taken into consideration in governmental policies. It is questionable that this has
been the case for regional development. Thus, while 1996 saw the publication of a
research programme on regional development from a gender perspective, and
while women as scholars have been able to gather and discuss this matter, to date
there has been no implementation of initiatives of this kind. Women so far have
been given space to argue about policy but their ideas have been put in a black box
(Briskin and Eliasson, 1999).

To understand the operations of the gender system, it should be recognized that it
is organized by both women and men, with women as well as men contributing to

this segregation. This might be one explanation of the gender divided Swedish labour market. Women have gained access to paid labour but women's participation in the labour market has been organized in accordance with women's responsibilities toward home and family. Even today, with high divorce rates, the heterosexual nuclear family is the standard societal agent beneath the welfare system. Hence, since benefits from the welfare system are connected to individual performance in the labour market, women's lower salaries are a structural problem (see also Haugen, 2004). Even more important, despite rhetoric on gender equality by the Government, is the way in which traditional family values seem to impact on those persons who define regional development policy. This might also explain women's more fragile relationship with the labour market compared to men. One indication of this is that there are employers who do not provide for women to take up full-time employment.

In effect, patriarchal societal structures still rule, which take men's lives and experiences as their organizing principles. This also impacts on what is considered to be important, real and needed in a regional development context. My starting point in this research was that women did not count when it comes to regional development. Regional development policy has been under transformation since the beginning of the 1990s. Before that, regional development policy and its subsidy system was about generating welfare in the whole country, focusing on industry to promote labour markets even in rural areas. At the end of the 1980s and at the beginning of the 1990s, policy turned toward competitiveness when the old policy was abandoned. Instead, all regions in Sweden are now expected to contribute to regional development by compete with each other.[3]

When regional development policy was formed by the mid 1960s, subsidies for industry were intended to neutralize geographical distances through transport subsidies, support companies to locate in remoter areas by giving locational subsidies and ease the costs for new employees by offering employment subsidies (Berger, 1995). Evaluations concluded that it was mainly men who benefited from these policies, through the companies that employed them. According to the Swedish policy of gender equality and the Swedish gender divided labour market, a quota system was connected to this specific subsidy. In one-sex dominated workplaces, the under-represented gender should be hired first (Olsson and Sundin, 1990). The 1990 evaluation of the gender quota system showed that the rule on the under-represented gender did have positive implications for women, although companies could still claim that women did not qualify for work within an industry (Olsson and Sundin, 1990). An evaluation 10 years later by Gunnel Forsberg (1999) states that the quota system has had a positive impact on gender mixed workplaces, even if there are regional variations depending on individual officials' decisions and regional gender contract constructions. This 'uncertainty' raises the question of whether women will gain from the European Union's regional funds in a specific way? This question is supported by a literature survey on regional development policy research reports, which concluded that gender had not been of particular interest in regional development policy (Lindkvist Scholten, 1996).

---

[3] There are still specific regions in need of direct aid from the government (*En politik för tillväxt och livskraft i hela landet*, 2001/02, p.4).

Time-Geography

In analyzing the actual experiences of women in seeking to establish businesses in Southeast Sweden, this study has explored actions through a time-geography framework. The time-geographical approach provides the tools that are needed for analyzing why the self-sustainable projects that the interviewed women tried to start did not work out. In time-geography, the connection between time and place is central, in an effort to understand processes and changes within society (Hägerstrand, 1991). This approach also makes it possible to relate the local and particular to the national and global. Time-geography uses the concepts of restrictions and resources associated with a project, which becomes important in analyzing successful as well as unsuccessful projects (Hägerstrand, 1985; Hallin, 1988; Friberg, 1990; Åquist, 1992). Time-geographical projects might be found on all societal levels, from cooking a dinner to negotiating inter-state affairs. About projects, the founder of time-geography, Torsten Hägerstrand (1985, p.201), states:

> To carry out a project requires that a succession of inputs have to be mobilized in an orderly fashion. An important feature of human plans is – as a rule – that they are not entirely rigid. Parts of a plan can change radically in terms of input, order, and location without violation of the overall purpose.

To carry out a project one has to gather knowledge and materials, and take control over a place where this project might be carried through. To be successful, one has to put all these different resources in their right positions at the right time. All projects might not succeed since they can compete for the same resources and some projects might be 'stronger' than others. Besides competition, there are three groups of restrictions, which impact on the outcome of projects. The first group of restrictions, capacity, is related to the individual. As biological organisms, people need to eat and sleep in specific cycles; our ability varies in different ways. The second group of restrictions is coupling. In order to prepare for a project, ideas are needed, along with the skill to carry the project forward and materials to carry the project through. The third group of restrictions is steering. This refers to systems of power on which the individual has little impact. Formal legislation offers one example of this. In my doctoral thesis, *Spaces for Women's Self-Support* (Scholten 2003), a fourth restriction, which relates to gender, was acknowledged. When analyzing my interviews, it became evident that different kinds of restriction interact. Gender, male industry traditions and the experiences of labour, resulted in a situation in which a woman interested in starting a business and business advisors were discussing two different types of business arrangement. The women I met in my interviews were mainly interested in livelihood businesses, arguing that they were living with someone with whom they shared their everyday costs. Running a business would give them an opportunity to develop something by their own hand that would be complementary to the household economy. To traditional business advisors, with a traditional approach to how to run a business, this would not count as a serious business attempt. In these examples, gender, coupling and steering restrictions co-work, making it hard for the individual to succeed in her attempt to become a businesswoman.

A Discursive Analytical Approach

Government texts on regional development policy, as well as texts at the local and regional government levels, have been important data sources for this study. Besides a traditional text analysis (Hellspong and Ledin, 1997), discourse analysis is used to focus on policy implementation. These kinds of texts are not ordinary text products, as they carry ideas within them that are transformed into policy, and are implemented by civil servants at local, regional and national governments levels into programmes of change (Fairclough, 1992). From a feminist perspective, it has been important to recognize who is speaking in these texts and whom these texts speak of (Mills, 1997; Said, 2002). Women are to some extent acknowledged, but women's experiences and everyday lives are not taken into consideration. These texts are built on traditional patriarchal values, based on a nuclear family organization, even if women are portrayed as being in charge of and responsible for their own lives.

To analyze regional development policy a hegemonic perspective might be helpful. Competition and growth are dominant 'discursive nodes' (Winther-Jörgensen and Phillips, 2000) in today's Swedish regional development policy (*Regional Tillväxt, För Arbete och Välfärd*, 1997/98, p.62; *En Politik För Tillväxt och Livskraft i Hela Landet*, 2001/02, p.4). Underneath these 'new' concepts traditional industrial policy is hidden. In order to make regions grow, population has to be acknowledged. Demography is important, because rural depopulation has left older people and less educated men behind. Governmental reports, and propositions, as well as regional government reports, have problems in addressing these issues. Sexuality, gender relationships, care and different forms of organizing family life are not defined as important questions to regional development policy. This is because the political Establishment has closed the discursive gate for alternative questions on how regional development might be understood (Fairclough, 1992; Winther-Jörgensen and Phillips, 2000).

When it comes to regional development documents and producers of regional development policy, documents play an important role in defining what is important and needed by society. These texts are the materialization of discourses on regional development, in which regional development policy focuses on large-scale structures (Berger, 1995; Scholten, 2003). Here public sector employees and representatives of regional partnerships create their own space of 'common' knowledge about regional development. If people with knowledge on gender do not participate in these discussions, and if gender is not viewed as a category of importance, then such texts will certainly not address questions of importance to women. Understandings on how gender is constructed and thereby positions women and men differently in society will not be challenged, nor will traditional gender roles. Yet regional development policy is supposed to be about how we might create sustainable societies over the whole country. With this aim, the translation of policy into programmes becomes an important eye of a needle in creating sustainable societies for both women and men.

Defenders of Hegemony

Participants in the established discourse tend to use gatekeepers in order to prevent interventions from 'intruders'. These gatekeepers are comprised of various groups of people, from civil servants to the managing directors of newspapers, each of which help decide who might participate in the discourse. For over 20 years, women as researchers and public investigators, who work with regional development policy have tried to make a difference.[4] Organizing symposiums, producing research programmes and literature on regional development and gender (Olsson and Sundin, 1990; Friberg, 1993; Tillväxt en fråga om kompetens, 1995; Statens Institut för Regional utveckling (SIR), 1996; 1996; Forsberg, 1999) are among the examples of what has happened. Those who represent the hegemonic regional development policy have paid this research and knowledge little attention. An important conclusion in *Spaces for Women's Self-Support* was that women are excluded from the discourse on what regional development could or should be about (Scholten, 2003). They are spoken about but they are not participating. An outcome of this is that a gender conscious perspective is lacking and women are defined as a homogenous group with similar interests (see Said, 2002, on discursive practices).

In the case of policies to encourage women's self-supporting strategies, there are organizations with the responsibility to implement Government policy. Besides unemployment offices, there are specific advisory organizations for businesses, one of which, ALMI, is responsible for administrating special loans and benefits for women who want to start their own business. This organization has come to play an important role in analyzing women's circumstances when they want to start their own business. ALMI is a corporation that brings together the interests of the County Council and NUTEK, which is one of the Government's agencies on regional development and business.

What my research has shown with regard to this critical organization for promoting women's interests in business formation is that 'man' is the norm that women have to adjust to. Fundamental to the ALMI worldview on businesses, there are no differences between men and women. What is critical are the conditions for business creation. These are the 'facts' that are examined when business proposals are put before ALMI:

> As advisor, you do not treat women and men differently. The business is judged on whether it is profitable or not and it has to have certain sales. It is strictly a business economic judgement. There cannot be any feelings involved in the process. (Business advisor, ALMI)

Indeed, the Corporation has analyzed what businesses it views as profitable to invest its money in:

> We have to ask ourselves in what way ALMI's resources may be of any benefit for the company. Our resources are supposed to be used in order to increase economic

---

[4]   Interview with Ulla Herlitz in 2003, reported in Scholten (2003).

value...what is primarily focused on is that women shall be able to withdraw payment from their company... ALMI doesn't refuse an application in order to be bloody-minded towards anyone, we act for the best of the customer. (Business advisor, ALMI)

What the advisors want, is that women, starting a business after long-term unemployment, have a business from which they can receive a salary from the first day. If the businesswoman is more interested in making a living, rather than creating a flourishing business, she is not motivated enough, according to these advisors. Advising men and women to start a business is a job that generates mixed emotions. Almost every one of the advisors interviewed talked about personal risks when starting a business of one's own. One of the advisors said that it would be an economic disaster for a person to end up with stock worth hundreds of thousands of krona and no business. What is really interesting in relation to those women I focused on in this research was that almost none were interested in large-scale ventures. One of the interviewed women started a joint-stock company in order to protect her assets, otherwise these women were not in charge of any real material goods. They did not want heavy bank loans. Indeed, when they needed a loan, they had problems securing one because they did not have the security the bank demanded.

Research into the business economy has reported on women's situations, and the terms they face when they want to start a business of their own (Nyberg, 1989; Sundin, 1989; Sundin and Holmquist, 1989; Holmquist and Sundin, 2002). The key message to come from this work is that there is no doubt that women face different conditions compared with men. Some of the advisors on business formation acknowledge the difficulty of women's situations:

To start a business of your own, you have to become like a man.

But for men on business boards, business advisors claim that:

Men talk about women as businesswomen, and as members in business boards, but women ought to be silent and docile. Women should not have their own ideas to accomplish. Women and their ideas are not treated with the same respect as men and men's ideas. (Business advisor, ALMI)

Indeed, amongst business advisors, the view was expressed that the problems faced by women who are in business are their own fault. One reason that is put forward to explain this is that women have not described or labelled themselves as businesswomen before. As a result of previous engagements with paid-work, women are seen to be responsible for their own harsh situation:

Hairdressers, women working as therapists and things like that did not call themselves businesswomen earlier... One of the great problems is that women tend to put their head in the sand. They just don't want to recognize problems... they just don't seem to be aware of conditions when running a business... their ideas on work are enormously traditional and women's paid-work has always been in low status jobs. (Business advisor, ALMI)

What is even more problematic from a feminist perspective is that these advisors maintain and create a traditional image of 'womanhood' as dependent, with values that are fully oriented toward the family and the vulnerable; that they are oriented as the 'other':

> Sometimes women motivate the budget for their business with arguments that 'I do not need that much, we have low costs, my husband is employed'. These are arguments a man never would put forward in a discussion on profit and the budget, telling me his wife has a good income and might provide for him… (Business advisor, ALMI)

Yet family circumstances are important here, for women living in family relations have to have the blessing of their family if they are to be successful. This became evident in a previous study (Lindkvist Scholten, 1999, 2001). Here one woman told me about family meetings, in which she tried to convince the rest of her family why she had seemed to become another person, which was partly because running a business is not something that can be done solely in office hours:

> To run a business is not to have office hours. It is a decision, which has to be taken with the family. Women have to have their husbands' support, otherwise it will be hard to be successful. It is a lifestyle to be a businesswoman. (Business advisor, NFC)

In this kind of context, starting a business as a response to hard conditions in local labour markets or as a result of entering unemployment is a response that interviewed business advisors were all sceptical about:

> Among businesses that are started out of long-term unemployment, at least 80 per cent shut down after six months. It is harsh but not negative. (Business advisor, NFC)

> …many of the companies started during the last period, [when] unemployed women have tried to make a living out of a business. Compared to women who weren't unemployed, the unemployed ones seek their outcome from the same [economic] branches. This is mainly the service sector, a branch [of the economy] that is put under serious competition. (Business advisor, ALMI)

Even so, there are business advisors who agree that a business run from home might be a way of balancing the different demands on women's lives. But it is important to run a business professionally. As one business advisor pointed out, you cannot use your private telephone, for what would a customer think if your five-year-old daughter answered the phone when s/he called?

Another interesting insight from the business advisors at ALMI is that they do not meet many women who want to start a business in the countryside. ALMI does not do jam-and-lemonade businesses, but works with companies they believe have growth potential. Women living in the countryside, I was told, are mostly interested in securing extra income on that gained from a farm business. The business advisors at ALMI are quite convinced that women living in rural villages or in the open countryside primarily turn to their County administrative board in order to obtain support from this source. There is also a special women's business advisor in those municipalities within the region that are defined as being in need of Structural Fund assistance, which the business advisors at ALMI can refer people to. A third set of organizations with an interest in regional development and women's businesses are the regional resource centres for women. These were not mentioned at all by the business advisors at ALMI as a potential source for assisting women with their businesses.

Battering-Rams

In the mid-1990s, the Swedish Government decided to put some effort into gender equality. A managing director of gender equality was employed by all county administration boards in Sweden, a national resource centre was established in Stockholm and regional resource centres were later recognized in 21 counties. In traditional regional development funding regions in the north, special women's business advisors were hired to encourage women to start their own businesses. This project was later extended to regional development regions in the south.

Both the resource centres for women, as well as women's business advisors, were projects initiated and supported by the Government. By the end of the project the aim was to establish these organizations within the fabric of local government. In some places these organizations have become internalized within the local government establishment but the focus is no longer women's businesses, but business development in general. The evaluations of the NUTEK project with women business advisors are consistent in declaring it a success (NUTEK, 1995, 1998a, 1998b, 2001). The women who worked as local developer and business advisors that I interviewed were grateful for this experience, even if they have doubts about the Government's aim in putting projects like these together in the first place (Scholten, 2003).

The main goal for women business advisors was to encourage women to start their own businesses. They had a free hand in organizing this work. In one sense they were seen as battering-rams that would break up traditional values concerning the role of women in business and labour, which was quite a hard and sometimes a thankless task.

> I'm speaking of men over 50, in some ways, they do not believe that we, women, have thinking capacity. I found out that they take us to be stupid, until you prove otherwise. You can't imagine all the attacks I have been under, when working as a women's business advisor. (Former women's business advisor)

The work of these women business advisors was primarily directed toward women living in rural areas, or at least those rural areas that were defined as being in need of structural change. Many women living in these areas, which were hit by economic recession, were older women, with less education and very often with work experience only from the public sector. These were not the clients ALMI referred to as potential businesswomen. Yet women's business advisors meet with enthusiasm and report that women are eager to create something of their own will. Characteristically, women's business advisors conclude that:

> You see, I meet all this gratitude, I feel really important! I have done so little to these women, so little, it is really a shame... (Former women's business advisor)

But what did they do to get this gratitude? In my interviews with former women's business advisors they regularly returned to the importance of talking, discussing and arguing from different angles, ideas on businesses and the totality of women's situations.

> What is perhaps most typical to our situation is talking. We talk a lot before addressing the hard issue. Counselling is never done once. One appointment takes about one and a half hours, no matter what we are talking about. (Former women's business advisor)

These business advisors were eager that no woman should start a business that would not be profitable. The effect of counselling was that the client had a lot of questions to think about after leaving a meeting. One important issue for one of the women's business advisors is to establish a positive self-image amongst women clients who are interested in starting their own businesses, trying to reach her clients' interests and create self-confidence within the single woman:

> Most important was to make sure they did the right thing, because many times I felt that the women did things others had told them to do... I tried to figure out what they would like to fulfil, deep inside their hearts. (Former women's business advisor)

Working for changed attitudes toward business and gender placed the women's business advisors in tough positions. They had to defend their counselling against other business advisors, for they were not considered to be professionals, even though they had run their own businesses. They experienced the paradox of being charged with changing local structures at the same time as they were part of the structures they were trying to change. One of the women's business advisors was located in the same building as a traditional business counselling office:

> When working with something these guys [the male business advisors at ALMI for instance] experience as threats towards their own advising, well, not threat, but totally different from their own professional counselling, well, these guys did not understand why I should be doing something they had experience of. Sometimes I had bad feelings going to work in the morning.

It is hard to be in-between different structures and women living in the countryside are quite aware of what structures they face if they decide to break informal societal rules, like becoming a breadwinner with a business of their own:

> There are norms, values and behaviour in rural areas, which cannot be ignored. These structures are old and marked by the church or perhaps other things. If you want to see things from a different perspective, you become odd, strange and you become looked at over one's shoulder. People wonder, what is she up to? You might think you have ended up in the dark ages! If you are a woman, then it becomes really problematic – 'what is she up to? What does she think about herself! She is a real weird one, you see.' It takes a lot to be accepted after that. A lot of women

hesitate to start a business of their own. They know the rules and they will not put themselves into this problematic situation. (Former women's business advisor)

Businesses and growth are important features for regional development policy (*En Politik För Tillväxt och Livskraft i Hela Landet*, 2001/02, p.4). But when it comes to women living in rural areas who want to accomplish a sustainable livelihood, then business is defined or constructed as something else. Rural women's projects are not primarily viewed as important to regional development. In the region where this research was conducted, manufacturing industry is the dominant business sector. But women do not primarily start their businesses within this sector. Thereby, they are not considered 'real' business people.

> This invisibility... how are women supposed to find and think and be creative in a situation where they are not defined as a human beings, creative and important for what they are doing? There was no supporting culture to manage from within for women. (Former women's business advisor)

The situation has become even tougher today, according to interviewed women's business advisors and the project leaders of regional resource centres. No one would admit that a hostile attitude exists towards gender equality. The Government likewise has spoken plainly of the importance of a gender perspective in policy. Yet resistance persists:

> Today I experience silent resistance. From different positions, you are being opposed without an opposition. Things are just being ignored or words never spoken. Still, some things are not possible to fulfil. I say, 'you don't say anything!', well, I got the answer, 'I have never said I was against you!' The effect is that you might not proceed with your work. Those with power are not clear or evident in their policy. This is not a new phenomenon, it is about positions and hierarchies, and when you have reached a certain position it is all about power and you meet these kinds of structures. (Project leader at a regional resource centre for women)

Smashed Livelihood Projects: Women Reflect on Failures as Businesswomen

The Government has put a lot of effort into convincing women that they should start their own business. To some extent these programmes have been successful. but that success does vary, depending on the level of education of the women involved, on what kind of businesses women start and, importantly from a geographical perspective, on where these businesses are established. Highly educated women in urbanized areas who are starting consultancy businesses have been the most successful group. Women in the countryside with low formal education who are interested in businesses in the recreation and handicraft sectors have been less successful (interview with the Managing Director of AMV, the Swedish National Labour Market Administration).[5]

---

[5] In the television society magazine *Reportrarna – de dyra jobben* SVT 29 June 1999 (Expensive jobs, the failures of Swedish national labour market administration).

The individual women who have been interviewed for this study live in rural areas. Some of them are among the less educated but there are women who have graduated from university. In interviews with these women I have met different arguments for why businesses did not succeed. These include: *place* – this was not the right place for me; *space* – there was not enough room for women's interests; *gender* connected to biological sex – if I had had a different appearance nobody might have argued with me; *gender stereotypes* – household responsibilities were mine as well; *time* – this was not the time for this kind of business; and, *kind of business* – I was told it would not be easy, but I am not afraid of hard work.

Erica, a young mother of two school-aged children, ran her business for two years. She guided tours on horseback combined with night school classes and selling minerals and vitamins. She started her business because both she and her husband had become unemployed in the mid-1990s. Her husband became depressed. Erica, however, could not wait for a job to be offered her. In their residential area there are interesting places to visit. Erica had for a long time been interested in horses. The idea was to offer tourists guided tours on horseback in beautiful environments. She did not have to advertise, for the rumour spread, and in the area where she lives lots of German and Danish people are summer residents. Her horseback tours and her business flourished. The problem was winter. In order to support herself and her family she ran about eight night classes a week. The situation became problematic. Her husband, when unemployed, had helped her with the administration of her business. But after a year and a half he got a job in local government, 25 kilometres from home. Erica then had to organize her administrative routines by herself. She also had to take responsibility for the children and the household. She explains why her husband did not take part in family responsibilities as partly to do with the strength of his job in pulling at him, but another explanation is that she ran her business from home, so she was seen to be in the home place, so she could run the household as well. Erika thought hard about what motherhood ought to be and the extent to which she lived to the cultural and traditional standards of being a mother. Some days she explained, she just yelled out orders to the children from the back of a horse, telling them what to do and what to eat. Then she would be off on her guided tours and the children were left on their own.

Nearly two years after the start of her business, the situation became intolerable. She had been saving letters from officials for several months and had a nervous breakdown. She was unable to run the administrative part of her business, and became afraid every time she received a letter from the local tax office. In the end, they had to sell all the horses and lost their home. Today she is grateful for the experience, and she has come to understand that the local tax office might provide help as well. She is fortunate that she never had any bank loans. The money she needed for expenses, her mother had lent her. At the same time she feels very angry with the unemployment office which, she claims, never told her about all the difficulties women and men have to overcome who want a business to succeed. Today, she would like to become a business 'dissuader', telling people not to start their own business and what they might end-up with if they do.

The second example I should like to present is that of Liza. In interview, Liza told me that beforehand she could have been dissuaded from starting a business, because she never could figure out the options involved. Yet, compared to Erica, Liza had graduated from university, and the task of answering letters and filling in forms created no problems. Still, circumstances led her to take a situation as an unpaid stablewoman. Her problem was to start a business as a woman in a particularly male branch of the rural economy. Liza had for long time been interested in horses, and she was a horseback rider. Her business was organized around the training and treatment of (horse) trotters. In order to secure her assets, business advisors told her that she should start her business as an incorporation.

The idea behind her starting her own business was that she would be able to run it by the ideals and norms that were important to her. She was tired of businesses that were incorrect in her opinion. Still, in order to run this kind of business, it is useful to be located near other businesses of a similar kind, because you sometimes need help with difficult horses. Waiting for her own place to be put in order, she hired stable places at her former employer's place. She described her situation when potential customers arrived. She was set aside, because she was a woman. Her colleagues often told her that she would not be able to handle particular horses. These men, running the business where she initially started her own company, had been without competition for a long time, and a woman had never before challenged them in their own backyard. Still, Liza was convinced that her business would be a success. She found her own place, with a trot-track and necessary equipment. The contract she signed for her business with the owner of the stable said she would be in business in a year. There was, however, a lot to be done in order to train the horses. According to the contract the owner was responsible for renovating the stable and track. Days became weeks and weeks turned into months. The owner of the stable had drinking problems and the contract had no options if it was not fulfilled. The day for the opening was close when she realized her contract was useless and her business would not open as advertised in the paper.

Her former colleagues started to became hostile towards her. She was met with comments on her appearance, on her dress behaviour, and on her business ideas. When she was advertising for her business, men offered to turn themselves in for a massage instead of their horses. Her business advisor at ALMI had given her a cap with the words, 'I'm the boss'. He knew what it would be like, which Liza said, 'I could never imagine!' Her business ended with abuse. A former colleague hit her and she left with a horse and a dog. Today, she supports herself as a personal instructor and does not have the strength to think about a business of her own. She never took any bank loans for her business. She had a business scholarship and was told by the business advisor that she might get a loan if she applied, because her idea was solid, she had the skill, the knowledge and even important contacts.

These two examples are both on horses, but Erica's business addressed private individuals interested in horseback experiences in beautiful surroundings. Liza, by contrast, was offering services to other horse trainers and companies, even if she also accepted private horses. Erica had a private firm and Liza was told to run an incorporation. The two businesses were in fact quite different. Why did Erica and Liza experience the same smashed livelihood strategies?

Space for Women's Self-Supporting Strategies in Regional Development: A Feminist Conclusion

Gender has always been of importance to regional development but constructed as a gender-neutral policy, administrators, business leaders and politicians have foreseen that regional policy has been directed towards men's labour markets. Changing regional development policy into a growth race between Swedish regions may result in competition between public and private sectors and in the long run between women and men's labour markets. Enlarging regions in order to become more competitive has negative effects on women, as women still have the major responsibility for children and the household. Instead, local solutions to specific problems have to be acknowledged as a way to promote regional development. Research has shown that women are concerned about local and regional development (Friberg, 1996), but whereas policy-makers are interested in international airports, women want cheaper, better quality local and regional transportation.

Another important issue that has to be addressed is livelihood strategies and self-reliance. According to the National Social Insurance Board, more than 500 000 individuals have retired early from work.[6] In the northern parts of the country, social insurance is an important part of livelihood strategies.[7] The balance between providers and supported asks for new solutions. The women interviewed in this study are not considered businesswomen according to the norms and values within male dominated contemporary business environment. Interested in labour market issues and regional development, we have to ask ourselves whether the norm is correct or not. Comparing Erica and Liza once more, Erica met with no obstructions to business establishment, because she organized her business at the margins of the local economy. Liza, by contrast, tried to establish within the local economy in a way women are not supposed to do and was met by abuse. If women organize their businesses in the in-between 'spaces' of local economies, they might be left alone. The problem with organizing on the margins is that you are not considered a businesswoman and your activity is not categorized as 'real' business.

Swedish society has come far in terms of formal legislation on gender equality. At the national government level women and gender equality are acknowledged in circumstances where policy is outlined for regional development, but regional partnerships within the regional development field are troublesome. The question is, when are women going to participate on the same terms as men do, and when will issues addressed by women have the same status within regional development as economic growth and industrial policy? Another problem is the dictatorship of numbers. Equal assemblies are important because of what this signals. But creating true equality is about challenging patriarchal norms and values. In order to change regional development and give opportunities to places outside urbanized areas, new conceptions of what is considered paid labour and a business have to be

---

6   http://www.rfv.se (24th November 2003).
7   Professor Einar Holm, Department of Geography, University of Umeå and Mr Lennart Sundberg, presented an ongoing study on social security benefits as livelihood strategies at the conference 'A Living Countryside', 1-2 September 2003, Östersund, Sweden.

acknowledged. The old political institutions need to let others, beside traditional value carriers, into the debating chambers on the future of our regions.[8]

References

Åquist A-C. (1992) *Tidsgeografi i Samspel med Samhällsteori*, Meddelanden från Lunds Universitets Geografiska Institution Avhandlingar 115, Lund.

Berger, S. (1995) *Samhällets Geografi*, NST [Nordisk Samhällsgeografisk Tidskrift], Uppsala.

Briskin, L. and Eliasson, M. (1999, eds.) *Women's Organizing and Public Policy in Canada and Sweden*, McGill-Queen's Universities Press, Montreal.

Bygder och Regioner i Utveckling, Governmental Report 1993/94, Fritze, Stockholm.

Cameron, B. and Gonäs, L. (1999) Women's response to economic and political integration in Canada and Sweden, in L. Briskin and M. Eliasson (eds.) *Women's Organizing and Public Policy in Canada and Sweden*, McGill-Queen's Universities Press, Montreal, pp.51-86.

County Administration Board of Jönköping (1996) *Mål 5b, Sydöstra Sverige 1996*, stencil, Jönköping.

Det Här Är Bara Början, NUTEK och ALMI, 2002, Stockholm.

En Politik För Tillväxt och Livskraft i Hela Landet, Governmental Report 2001/02, Fritze, Stockholm.

Fairclough, N. (1992) *Discourse and Social Change*, Polity, Cambridge.

Forsberg, G. (1999) *Könskvotering i Regionalpolitiken – Förslag till Framtida åtgärder*, IM-Gruppen Rapport 1999:1, Uppsala.

Friberg, T. (1990) *Kvinnors Vardag – Om Kvinnors Arbete och Liv, Anpassningsstrategier i Tid och Rum*, Meddelanden från Lunds Universitets Geografiska Institutioner Avhandlingar 109, Lund.

Friberg, T. (1993) *Den Andra Sidan Av Myntet – Om Regionalpolitikens Enögdhet*, Glesbygdsmyndigheten, Östersund.

Friberg, T. (1996) Ett könsteoretiskt perspektiv på regional identitet, in M. Idvall and A. Salomonsson (eds.) *Att Skapa en Region – Om Identitet och Territorium*, NORDrefo, Stockholm, pp.60-79.

Hallin, P-O. (1988) *Tid För Omställning: Om Hushållens Anpassningsstrategier Vid En Förändrad Energisituation*, Meddelanden från Lunds Universitets Geografiska Institution Avhandlingar 105, Lund.

Haugen, M.S. (2004) Rural women's employment opportunities and constraints: The Norwegian case, in H. Buller and K. Hoggart (eds.) *Women in the European Countryside*, Ashgate, Aldershot, 59-82.

Hellspong, L. and Ledin, P. (1997) *Vägar Genom Texten: Handbok i Brukstextanalys*, Studentlitteratur, Lund.

Hirdman, Y. (1989) *Att Lägga Livet Till Rätta: Studier i Svensk Folkhemspolitik*, Carlssons, Stockholm.

Hirdman, Y. (1990) Om genussystemet, in *SOU 1990:44 Demokrati och Makt i Sverige, Maktutredningens Huvudrapport*, Allmänna förlaget, Stockholm.

Hirdman, Y. (2001) *Genus - Om Det Stabilas Föränderliga Former*, Liber, Malmö.

Holmquist, C. and Sundin, E. (2002, eds.) *Företagerskan – Om Kvinnor och Entreprenörskap*, SNS Förlag, Stockholm.

---

[8] I should like to express my gratitude to Keith Hoggart for encouragement in producing this chapter and for helpful comments on the text.

Hägerstrand, T. (1985) Time-geography: focus on the corporeality of man, society and environment, in *The Science and Praxis of Complexity*, United Nations University, Tokyo, pp.193-216.

Hägerstrand, T. (1991) Tidsgeografi, in G. Carlestam and B. Sollbe (eds.) *Om Tidens Vidd och Tingens Ordning*, Byggforskningsrådet T:21, Stockholm, pp.133-142.

Lindkvist Scholten, C. (1996) *Tankar Kring Kvinnor och Regional Utveckling*, Appendix to Kvinnor och män i dialog om regionernas framtid, stencil, Statens Institut för Regional utveckling, Östersund.

Lindkvist Scholten, C. (1999) *Kvinnors Försörjning på Landsbygd*, Meddelanden från Lunds Universitets Geografiska Institutioner, Licentiatavhandling, Lund.

Lindkvist Scholten, C. (2001) Place, gender and strategies for self-support, *Geography Research Forum*, 21, pp.15-29.

Massey, D.B. (1994) *Space, Place and Gender*, Polity, Oxford.

Mills, S. (1997) *Discourse*, Routledge, London.

NUTEK R (1996) *Affärsrådgivare För Kvinnor: Att Främja Kvinnors Företagande*. Slutrapport, Stockholm.

NUTEK R (1998a) *Affärsrådgivare För Kvinnor*, utvärdering del 1. Förutsättningar för förändring, Stockholm.

NUTEK R (1998b) *Affärsrådgivare För Kvinnor*, utvärdering del 2. Identitetsskapande inom lokalt näringslivsarbete, Stockholm.

NUTEK, R (2001) *Affärsrådgivare För Kvinnor, Slutrapport Från Projektets Utvecklingsförlopp i Södra Sverige*, Stockholm.

Nyberg, A. (1989) *Tekniken - Kvinnornas Befriare? Hushållsteknik, Köpevaror, Gifta Kvinnors Hushållsarbetstid och Förvärvsdeltagande 1930-talet - 1980-talet*, Linköpings Universitet, Linköping.

Olsson, B. and Sundin, E. (1990) *Könskvotering i Regionalpolitiken, Expertgruppen för forskning om regional utveckling Ds 1990:54*, Allmänna förlaget, Stockholm.

Said, E. W. (2002) *Orientalism*, Ordfront, Stockholm.

Scholten, C. (2003) *Kvinnors Försörjningsrum –Hegemonins Förvaltare och Murbräckor*, Meddelanden från Lunds Universitets Geografiska Institution Avhandlingar 149, Lund.

SOU (1963) *Aktiv lokaliseringspolitik*, Esselte, Stockholm, p.58.

SOU (2000) *På väg mot nya arbetssätt för regional utveckling. Regionalpolitiska utredningens slutbetänkande*, Fritzes, Stockholm.

Statens Institut för Regional utveckling (SIR) (1996) *Kvinnor och män i dialog om regionernas framtid*, Östersund.

Regional Tillväxt, För Arbete och Välfärd, Governmental Report 1997/98, Riksdagens tryckeriexpedition, Stockholm.

Sundin, E. (1989) Kvinnors liv och arbete och regional utveckling – en mångtydig bild, in Birgit Helene Jevnaker (ed.) *Kvinnor och Regional utveckling*, NordREFO, Borgå, pp.68-84.

Sundin, E. and Holmquist, C. (1989) *Kvinnor Som Företagare: Osynlighet, Mångfald, Anpassning*, Liber, Malmö.

Tillväxt – En Fråga Om Kompetens, Meddelande, Länsstyrelsen, Västerbottens län 1995, Umeå.

Walby, S. (1990) *Theorizing Patriarchy*, Blackwell, Oxford.

Winther-Jørgensen, M. and Phillips, L. (2000) *Diskursanalyse Som Teori og Metode*, Samfundslitteratur Roskilde Universitets Forlag, Fredriksberg.

Chapter 7

Rural Women in the Former GDR: A Generation Lost?

Bettina van Hoven

Introduction

Since German unification, women in rural areas of the former GDR have experienced numerous changes. The implementation of a new political system created considerable space for previously absent freedoms and consumer choices. 'Everything was still possible' (029)[1] and then-Chancellor Kohl promised Eastern Germany 'blooming landscapes'. However, the human cost of transition was high. Several studies of women in the former GDR have noted that women as a group were among the main groups of 'losers' from transition (Rueschemeyer, 1994; Shaw, 1996) and that new political ideologies have made it easier for men to discriminate against women (Graham and Regulska, 1997). Consequently, women were ascribed an inferior role in post-socialist societies. This translated into a reduction or loss of childcare facilities, restrictions on birth-control, discrimination in the labour market and a return to a more traditional role for women as mother and housewife.

Some regions such as South Thuringia, near the West German border, have since begun to recover and women have taken on new jobs and responsibilities, as well as regaining a sense of empowerment (Pfaffenbach, 2002). Overall, the Amsterdam Treaty of May 1999 influenced political thinking about gender issues and the implementation of measures to facilitate the equality of men and women in political, social and economic issues was promoted. However, in remote areas such as in the eastern parts of Mecklenburg-Westpommerania (in particular the region of Uecker Randow), political and economic change caused many women to feel unsettled, displaced, isolated and 'useless'. In particular, the effect of widespread and long-term unemployment on women's everyday lives has been dramatic. Almost 15 years after unification, a significant number of women who are older

---

[1] The numbers in parentheses indicate respondents who contributed to the research through correspondence. Respondents in focus groups are identified by a fictional first name.

than 45 years remain unable to integrate into the 'new' system. New policies, generated as a result of the Amsterdam Treaty, have had no impact on the lives of these women and some policy-makers have begun to talk about a 'lost generation'.

This chapter draws on research for my doctoral research that was conducted between 1996 and 1999, along with a follow-up study in 2002, on women's changing geographies in Mecklenburg-Westpommerania since German unification (van Hoven-Iganski, 2000). Starting in October 1996, I corresponded with 40 women in Mecklenburg-Westpommerania about their experiences around the time of unification and thereafter. During a later phase in the research, in-depth interviews were conducted with key informants representing various regional and local political and non-governmental groups. In addition, focus groups were held in six remote villages near the Polish border. Of the correspondents and focus group participants, most women were more than 40 years old and had worked (and lived) within the agricultural sector all of their lives. Since unification, few had changed careers but a significant number of women had experience in one of the short-term government-funded employment schemes. Many were had retired early or were without work.

In this chapter the experiences of these women are examined, drawing out particular dimensions of women's lives that were affected by the transition, focusing predominantly on impacts on working lives, but also on communal life and engagement in local politics. This discussion will be preceded by a recollection of unification in the GDR in general and in the study area in particular. Throughout, experiences of the present will alternate with experiences of the past in order to compare and contrast what it was like for these women to live through the transition.

Towards an 'Overnight' Unification

The image many people may have of German unification is one of East Germans conquering the Berlin Wall by climbing on it, cheering from its top and eventually breaking it down. Preceding this symbolic event was, of course, a time of change in the former GDR and elsewhere in the former 'Eastern Bloc'. Mikhail Gorbachev, the President of the Soviet Union in the mid-1980s, initiated reforms toward the liberalization of socialist societies and economies. Interest groups began to gain a public voice in supporting reforms from below. In the GDR, the New Forum roundtable discussions and the Monday Demonstrations in Leipzig were signals of increasing public demand for political change. At the same time, the collapse of the GDR regime was accelerated by the fact that Hungary opened its border to Austria. More than 65 000 GDR citizens used this opportunity to flee to the West between August and October 1989. Then Chancellor Kohl interpreted these signs as a nation-wide need for unification and, within one month after the fall of the Berlin Wall, prepared a programme for the establishment of a German Federation that included the territory of the former GDR.

Although a transition period of 10 years of equal partnership was initially envisaged, the unification process was soon expedited (Conradt, 1996). As a result, little time was devoted to exploring and evaluating the democratic efforts by East German interest groups. Little attempt was made to involve these groups in the development of a 'New Germany'. It is notable that these groups had not favoured German unification but preferred change within the GDR. They had called for an alternative to capitalism (Hudelson, 1993), a 'third way' that would combine the virtues of socialism and capitalism (Stark, 1996). According to them, positive aspects of the socialist system included the right to work and housing, free health care and social policies that enabled women to combine work and family duties. Key benefits associated with the West were related to new freedoms. Although the activities and discussions of interest groups had been characteristic of the pre-unification period, the lack of political experience and organization of these groups eventually led to their marginalization and absorption into dominant Western political procedures. The possibility of negotiating a new constitution approved by a people's referendum was neglected in favour of a 'swift' unification. As a result, both the legal framework and key political actors were 'imported' from the West, evoking a sense of colonization among many East Germans (Wollmann, 1995).

The active involvement of urban GDR citizens in the process of unification through forums, protests and interest groups was not uniform. In more remote and rural areas, life continued more or less as always. For many people in these areas, the events that were to come were difficult to place and, equally, they found it difficult to place themselves in the 'New Germany'. Uecker Randow, a region in the northeast of Eastern Germany, near the Polish border, is one of these regions. It makes sense to look back at how this region developed under socialism and to introduce what life for women was like in order to understand how women in this region experienced unification and life thereafter.

Life in Uecker Randow in the Socialist Era

In an official promotional publication by the government office, Mecklenburg-Westpommerania is described as being located 'strategically well ... at the seam between middle, north and eastern Europe' (Staatskanzlei, 1997, p.6; Figure 7.1). Inhabitants of this region, however, characterize it as a place 'behind the seven hills' (referring to the remote homes of Snow White's seven dwarfs) or a 'black hole'. Marked by vast agricultural areas, almost two-thirds of the total area is arable. But in Uecker Randow soils are poor, nature is vast and industry scarce. Population densities have been low, with often less than 20 people per square kilometre. Until unification, infrastructure was poorly developed and roads poorly maintained. Under socialism, most villages were typically 'socialist villages' (Herrenknecht, 1995). This implies more than that they were residential (and working) areas under the socialist regime. Indeed much effort had gone into carefully planning the layout of villages in terms of size and function and planning the working and social lives of each individual.

Figure 7.1 Location of Mecklenburg-Westpommerania within Germany

The LPG as a Village

After the Second World War, large and medium size landowners, and those who were allegedly Nazi supporters, were expropriated (Reichelt, 1992). Until the 1980s, the socialist regime turned this property into large-scale industry-like structures in the form of agricultural co-operatives (Landwirtschaftliche Produktionsgenossenschaften, hereafter LPGs). In particular in the 1970s and 1980s these co-operatives underwent specialization, intensification and the concentration of production (Vogeler, 1996). They were subsumed under State control, enabling the government to exert significant influence over production plans and the appointment of key personnel at the local level. The creation of large-scale agricultural co-operatives concentrated enterprises and their employees into a small number of locations, thereby creating regional mono-structures.

The LPG was commonly regarded as the political, economic, cultural and communal centre, and agriculture was the principal force for innovation in their associated village. The LPG was equivalent to the village in terms of its boundaries and its authority relations. Hence, all changes in LPG structures impacted on development in villages. Targets for development in villages were regulated by means of yearly plans, such as social, cultural and women's furtherance plans, all of which were a part of LPG plans.[2] The LPG was, furthermore, responsible for the maintenance of built structures, road repairs and housing stock (Beer and Müller, 1993; DeSoto and Panzig, 1994; Vogeler, 1996). As one interviewed mayor concluded, 'the LPG was the A-Z of the village'.

The LPG played an important role as an instrument in assisting the State to raise the status of co-operative farmers to that of industrial workers and to achieve full social security, social equality and above average incomes. A sense of community and solidarity were characteristic values that were to be established amongst co-operative farmers (Watzek, 1997). Consequently, socialist and farming traditions, such as seasonal celebrations, were revived and reinforced. Celebrations included the 'International Women's Day', the 'Day of the Republic', summer, harvest and Christmas festivities, 'Children's Day', 1 May and several festivities organized by groups such as the youth and women's organization or even by individuals (i.e. weddings, anniversaries). Preparations for these events were mostly carried out by women in addition to their usual responsibilities. This level of participation was, in part, a consequence of political and social pressure. It was a part of 'socially useful work', that is, 'voluntary' work in which the majority of co-operative workers were involved. Positive engagements were rewarded by awards (Shaffer, 1981) and by a small financial incentive.

Women's Politics

An important part of life in a socialist village evolved around work. Although the LPG had many social functions it was primarily a place to work. In the GDR, work-related policies, such as the right to work, were at the core of the political system. As a result, the vast majority of GDR citizens were gainfully employed, with differences based on social class abolished and downward social mobility virtually impossible. In particular from the 1960s onwards, a set of policies was devised specifically targeting the integration of women into the labour process. These policies, the so-called *Frauenpolitik*, also addressed women's education and political engagements. These efforts were quite successful as the level and number of qualified women increased throughout socialism. In 1985, 48 per cent of all skilled workers, 12 per cent of supervisors, 62 per cent of all polytechnic graduates and 38 per cent of all university graduates were women (Belwe, 1989). By the end of the 1980s, 90 per cent of women were integrated into the labour market. Although a number of women were placed in stereotypical male jobs (DeSoto and

---

[2] Women's furtherance plans were an element of positive discrimination in favour of women and included detailed plans for the achievement of women's qualifications, social activities and socio-political education, for instance.

Panzig, 1994), for example as tractor drivers, women's employment was nevertheless concentrated largely in occupations in the service sector, in education, in public health and in the social services; all of which are sectors with lower paid jobs. Women represented 72.2 per cent, 77.0 per cent, 83.3 per cent and 91.8 per cent of the workforce in these sectors, respectively (Nickel, 1990a, 1990b; Marx-Feree, 1993). In particular, in the agricultural sector, women were represented predominantly in manual, labour-intensive work rather than occupying leadership or managerial positions. It is worth noting, though, that although women's income was on average lower than that for men, their contribution to total household income was 40 per cent (Marx-Ferree, 1993).

As I noted above, women were actively involved in the organization of festivities in the LPG. Many activities were run by women's committees, which were established at the workplace and at the communal level in order to increase women's political awareness (Clemens, 1990; McCauley, 1983). Weekly discussions addressed issues such as the opening hours of local shops and childcare facilities, the contents of women's furtherance plans at the workplace and social events. Although a number of women felt that they influenced local developments, it must be stressed that women did not usually participate in official party politics. However, women's involvement was higher at the local and regional levels. Even so, in party politics, as members of the Socialist Unity Party (SED), women were poorly represented, with just 35.5 per cent of all members in 1985. Women's representation at the highest level, namely the Central Committee, was but 13.5 per cent, none of whom were full members of the 'Politbüro' (Einhorn, 1989; Rueschemeyer and Szelény, 1989).

As women became more active in employment, birth rates began to drop. As a result, political propaganda began to emphasize the role of children as central to women's lives and women were encouraged to combine motherhood with employment. A series of measures were implemented that were largely childcare-related. These included extended paid leave after each child with a guaranteed return to a workplace, free contraception, a state birth premium of 1,000 Marks, extensive childcare services and, for young couples, generous low-interest loans (Marx-Ferree, 1993; Kolinsky, 1996). A number of domestic responsibilities were moved into the public realm, such as laundry services, mending services and canteens. The opening hours of childcare facilities and local shops were organized around women's work times and many women had the opportunity to do their shopping, or pay a visit to the doctor or hairdresser, during work-time. As most services were provided locally, there was little need to be mobile. The majority of trips to social or political events elsewhere were organized by the employer, who also provided transportation.

The above description implies a high level of state influence on women through LPGs. The political agenda of the socialist State indeed infiltrated almost every facet of their lives. However, it appears that, at least in the villages studied, women adapted to these pressures and found comfort in the social structures within their collective.

Experiencing the Wende in Uecker Randow

Most women respondents in the studied region were politically indifferent and lived on a basis that implied 'as much as necessary and as little as possible' political involvement. It is notable that this attitude has its roots in legacies of the past. In contrast with many areas near cities or in the south of the GDR, regional development in the district of Neubrandenburg (of which Uecker Randow was a part then) throughout the 1950s and 1960s was not significantly marked by social upheaval. The revolt in Berlin in 1953, the protests in Poland in 1955, in Hungary in 1956, and in Czechoslovakia in 1968 (Conradt, 1996) remained at some distance from farmers in this district. Although political change in the GDR was widely anticipated and desired, unification itself presented a shock to many East Germans. In Uecker Randow, memories of unification were certainly less vibrant than events in cities would suggest. Thus, Doris, an accountant in an agricultural co-operative, remembers the night when the Berlin Wall came down:

> It was just there, overnight. We had an assembly here and then Peter came, he was the last one [and arrived] half hour late. And he came and said 'The border is open', and we said 'You're nuts!'

Elsewhere, similar descriptions were given. A former mayor illustrates:

> In towns, surely it was different. Maybe because there are more people. Or because they worked so much here, did so much overtime. That's possible, too. I mean, who did occupy oneself with other stuff? Nobody. It was more or less hunky-dory. [...] On the 'Day of the Republic' in October we had a proper celebration and everyone was invited [...] We had a blast! Everyone was there. We [were] ... together... with a large core [of the LPG] and celebrated October 7th, really superb and snug.

Although the above quotes indicate an element of surprise, this is not to say that events in cities and elsewhere in the 'Eastern Bloc' had gone unnoticed. As many people were unable to receive Western television, much news reporting was passed on by family members or friends in Poland or West Germany. The following conversation reflects the feelings some women had about these events:

> *Karla*: Isn't it funny that they already knew in Poland? We have relatives and they told us that Germany will become one Germany. They already told us before we knew.

> *Monika*: I must be honest, when it all started and [those people] left for Hungary and the Czech Republic and so [on], I thought 'My God, they're so stupid!' Especially if you saw that with those babies [that were left behind] and so [on]. I thought over and over again 'They are stupid. *What* are they doing?'

Some villagers had gone to explore the West and most villagers anticipated the introduction of Western consumer goods. But many respondents recalled sobering experiences, which prevented them from wanting to live in the West. As one correspondent wrote:

Aside from the glittering wrap, we also saw much suffering and poverty (beggars in front of a department store being chased away by the police). I got to know people with a second home on Mallorca, a Mercedes for every member of the family and travel abroad twice a year... But I also met people who lived off social welfare with no extras. I saw absolute waste and, at the same time, incredible poverty. (064)

The events to come were equally sobering. Rather than experiencing the merits of Western capitalism, rural areas, and particularly small villages, needed to come to terms first with the process of agricultural restructuring.

Agricultural Restructuring

At the time of unification, there were approximately 270 LPGs cultivating more than two-thirds of arable land in Mecklenburg-Westpommerania. In order to regain private ownership from co-operatives, the new federal State initiated extensive 'land readjustments'. The aim of land readjustment was '... the development of a multi-faceted agricultural structure and the creation of suitable conditions for the re-establishment of competitive enterprises to enable participation of all people, who are occupied in the sector of agriculture, in the development of incomes and prosperity' (LwAnpG, 1996, §3). In effect, this meant that farmers who had been disowned between 1933 and 1945 and those after 1949 until 1960 (see, for instance, Merl, 1991; Smith, 1996), many of whom had migrated to the West, could reclaim their land.

In order to develop 'suitable conditions' for the market economy, government schemes were devised to encourage farmers to start family farms, or establish new types of co-operatives (GmbH, AG or GbR). However, setting up new enterprises was largely inhibited by a lack of capital as well as undeveloped marketing knowledge and experience. In addition to reticent East German entrepreneurship, a number of West Germans, previously the 'capitalist enemy', successfully reclaimed arable land and property, and began establishing individual farms. In so doing, it was predominantly West Germans who benefited from government grants and bureaucratic experience. Stark (1996) gave an example of this concerning the distribution of Treuhand properties, whereby 90 per cent of the privatized firms were sold to West Germans. Consequently, East German farmers felt they had been subject to a second expropriation. Indeed, most respondents in the research for this chapter spoke of being colonized by the West.

Ignoring social problems related to the reclamation of property, at the time of this study, the agricultural restructuring process had largely been accomplished in East Germany and those farms that survived or were newly established operate within the structures of the Common Agricultural Policy of the European Community. But there were variations in how well the villages managed to co-ordinate the restructuring process economically with emerging socio-cultural problems. Communities had to find their own strategies when attempting to overcome old political structures and to institutionalize new administrative ones. For example, the success of privatization of local facilities formerly associated with the LPGs had a direct impact on the restructuring of the social composition of rural communities, because the local labour market, the economic and social 'wealth' of the village, and the continuity of community services, were all affected. It is important to note that the loss of jobs within the former LPG was significant in small villages, such as those studied, since, in the majority of these villages, no alternative employment opportunities existed (especially for women). Most work was provided through government funded short-term employment measures.

Women's Integration into the Labour Market Today

In an attempt to cushion the impact of unemployment after unification, the government introduced a range of early retirement and short-term employment schemes. Although in 1992, 68 per cent of all unemployed persons in Mecklenburg-Westpommerania were women (Boje *et al.*, 1992), gendered patterns were consolidated as women had unequal access to these governmental schemes. Thus, while early retirement could be granted to workers between 55 and 65 years, some 70 per cent of women over 55 years took early retirement, compared with 50 per cent of men. The following key reasons accounted for women's disadvantaged position in the labour market (Fink and Grajewski, 1994):

- Women are made redundant more frequently than men (and men are twice as likely to find work again; Schumann and Jahn, 1991);
- Higher obligations to the family result in more difficulty in securing re-entry into the labour market (90 per cent of women are mothers and 30 per cent are single mothers (Nickel, 1990a));
- Low transport mobility; and
- Fewer employment opportunities for women through short-term, subsidized employment measures.

More recent figures demonstrate that the above trend has continued. For example, a report by the government of Mecklenburg-Westpommerania noted that, overall, more women than men were unemployed (Landesregierung Mecklenburg-Vorpommern, 2000). In April 2000, 50.6 per cent of the unemployed were women. However, the employment agencies have only been able to help 42.9 per cent of women to find a job in the 'primary labour market'. In contrast, more women are at present in the 'secondary labour market', that is, in short-term government-funded employment, in further education or retraining. In April 2000, the share of women in short-term government-funded employment (ABM) was 57.0 per cent and their share in further education was 52.5 per cent. Although many younger women have higher qualifications than men, most of them choose one of 15 gender-stereotyped (female) jobs.[3] In addition, many young women aged between 15 and 30 years have begun to migrate to the West. More than two-thirds of all immigrants fall into this group.

---

[3] As an illustration of such gender-stereotyped jobs, more than 90 per cent of the workforce are women in the garment and textile industry, as they are in the postal service, with this percentage also holding for salespersons and secretaries.

This is perhaps not surprising, as the average income of female employees in 2000 was 65-71 per cent of that earned by males. This translates into the following distribution: 79 per cent of employed women received less than 1,100 euro per month (net income), approximately 40 per cent of employed women earn less than 700 euro per month and only 3-8 per cent of female managers are represented at higher decision-making levels. In small firms, the share of female managers is about 7 per cent whilst in businesses with more than 5 000 employees the figure is reduced to 3.4 per cent.

Providing further evidence of the inferior economic position of women, more than 80 per cent of women in Mecklenburg-Westpommerania work part-time. Moreover, of the long-term unemployed, two-thirds are women. In particular single mothers are at risk of dropping below minimum income levels. More than 40 per cent of single parents have a yearly income of less than 9 000 euro and more than 50 per cent of single parents already live off welfare. In the longer run, women are also disadvantaged once they become pensioners. The German pension system is based on the work biography of an average male. Women, with their lower incomes, high levels of part-time work and likely employment interruptions due to childbearing and –rearing, receive less than two-thirds of an average male pension.

The Impact of Unemployment on Communal Life

Initially after unification, the dramatic rise in unemployment and the vast reduction in community services and health care were problematic for the continuity of social life. This process led to enhanced social fragmentation, pitting the unemployed against the employed, the old versus the young, new residents versus old residents, mobile persons versus immobile persons, people with a politically ambiguous history versus those with none and, to some degree, men versus women. Commonly, villages were characterized by the presence of the unemployed, pensioners and children, all of whom can be characterized as residents without personal mobility. Since women comprised a large number of unemployed and carers for small children and elderly people, they were more likely to be confined to the village than men. The working population (largely men) needed to be mobile in order to accept work at some distance from their home. This distinction between the employed and unemployed led to differentiated access to social services and goods. In the following sections, I will discuss these experiences from the perspective of a group of unemployed women.

The Meaning of Work and Unemployment

The meaning of unemployment for women's everyday lives must be seen in relation to their experiences of employment in the GDR. The interrelationship between the workplace, social life, and public participation in the GDR, as indicated above, must be strongly emphasized when attempting to understand past experiences. All elements overlapped significantly in women's everyday lives.

Social activities took place around or within the set framework of work but were never disconnected. Club activities took place after working hours but included colleagues and were sponsored by the LPG. Participation was essential if one aimed to obtain promotion, an award or a higher end-of-year premium. Social events took place in the context of the clubs or working collectives, with the LPG both organizing and sponsoring these events. Social events, such as holidays in a resort, were also awarded for good work performance. The resort itself was supervised by the workplace and maintained by co-operative workers. The family became an integral part of these practices as well. Partners were always invited to the social events listed above. Children were commonly involved in the workplace by means of 'treaties' between schools and the LPG. Furthermore, in their class, or as 'Pioneers', children would prepare little plays, dances or sing-alongs to be performed during the celebrations of LPG workers. Aside from these elements, education and further qualifications, health care and basic services were all associated with the LPG and thus with the workplace.

Paid-work in Western societies and in East Germany today has a less dramatic and less overarching function. Although work can establish links to social life, to public participation, and to the family, this would largely be a result of personal desire, ambition and initiative. It would most certainly not take place in a politically planned and fully sponsored way, organized by the employer, as was the case in the GDR. The respondents' accounts suggested that women did not consciously associate their previous social well-being with the complex interaction of various community services subsumed to the LPG. Therefore, they were not aware of how comprehensively their former workplace had permeated every part of their lives. When the transformation from socialism to a market economy shattered the old workplace relationships, former community functions were often geographically dispersed. As many women had regarded the work-place as a means to their self-identification (see van Hoven, 2001), the disintegration of it and the dispersion of its components constituted a considerable problem for women's adaptation to their new social environment and the formation of a new identity. Many women had expectations of about social life and public integration that were almost exclusively by means of employment.

A feeling of displacement from the local context and from the steady lifeline experienced in the GDR was a major theme in women's discussions of unemployment. Throughout the interviews, women emphasized that they felt listlessness, bitterness, desperation and depression. Reflecting on the meaning of work in the GDR and the social context of the workplace, it is notable that none of the women claimed to be relieved at being able to stay home and spend time with their families. Instead, they felt angry at being forced to 'play housewife' (011), to be 'back at the hearth' (010), or 'back with the pots' (017). Even though many women did not look for employment actively or write frequent job applications, as some key informants pointed out, they remained registered as unemployed not only to continue claiming benefits but to indicate that they were not content with their exclusion from the labour market. A number of women themselves indicated that the sharp decrease in birth rates since unification was 'proof' of women's choice of employment over family. This choice was supported by the absence of incentives to have children after unification. Whereas in 1989 almost two million children were born, in 1994 this figure had decreased by 60.4 per cent to 787 000 births. Although there was a slight increase again by 1996, the figure remained at less than 50 per cent of the 1989 birth rate (Statistisches Bundesamt, 1997).

But women's 'silent protest' was largely useless in terms of its efficiency. Many women remained trapped in a cycle of unemployment and short-term employment measures. Women's outlook on their future working biographies diminished. Whereas they first sought suitable employment, they began looking for *any* kind of job soon after unification. But at the time of their interviews most women were merely hoping for a short-term government funded job. Many women have also experienced some form of de-qualification since unification; for example, by taking work as fork-lift truck drivers when they have university degrees. However, few women respondents spoke of the fact that they were subject to discrimination, but instead felt guilty about their de-qualification and believed they were responsible for it themselves (similar conclusions were drawn by Eichener *et al.*, 1992). Because of the meaning women ascribed to the workplace prior to unification, women felt stigmatized and like 'good-for nothings'. They stated that 'inside, the soul gets ruined' (016) and that a 'tremendously important thing was taken away' (017) from them. Many respondents stated that there was no space for them any more and, therefore, they isolated themselves.

Throughout the discussions, few women referred to other people, such as family members or friends, as sources of comfort during their time of unemployment. Few women sought help when dealing with problems resulting from the new demands on their everyday lives. Their image of asking for support, for example, from unemployment agencies or family support agencies, was seen to be equivalent to going 'soliciting'. Some women felt that the information given to them, if they did look for advice, was not helpful, since the consultants themselves were only temporarily employed and were overwhelmed with the same problems they ought to solve for others. But when asked how women spent their time, most women claimed to be very busy doing things around the house or garden. Some women tried to connect with their former workplace but were either struck with their own sadness and desperation about their job loss, or felt that they were intruding on a space that was no longer theirs. Others described how they almost went out of their minds from being restless and helpless. The following excerpt from one respondent who was taking part in a group discussion illustrates some of these women's experiences:

Renate: The psychological consequences belong to it. We became unemployed immediately after unification ... we weren't paid much [in the GDR] and from that we got 63 per cent [unemployment benefit]. Meanwhile it's only unemployment support, that's 10 per cent less. And then staying home, one person bothers the other person ... one comes through one door, the other leaves. And one cannot really talk about the whole thing either... one has to cope with most of it on one's own, cope within, and nonetheless deal with all the rest. One has to budget and be careful not to be in debt... but, well, one always tried... I mean I just couldn't have coped at home, I must be honest. In the beginning, I always ran away from the house... into the garden and I cried. There was enough work in the house, but one just couldn't...

all alone... then one had to go somewhere to talk, to hear something else... I just couldn't cope.

Three women correspondents stated that they were so fearful, desperate and disillusioned after becoming unemployed that they considered committing suicide. This outcome was only averted with the help of close family and friends. Family was generally a prominent theme in women's writings, with many giving details about the importance of their family, 'for the soul' (065), and as an alternative to recreational activities outside of the house, or even to political participation.

Barriers to Political Participation

It is notable that all the activities that have been mentioned so far were confined to women's homes or the village. This is consistent with trends from the past, for women in the GDR have traditionally not had opportunities to travel much beyond their locality, and have little chance of taking up opportunities beyond their immediate vicinity. Even since unification, with limited governmental or outside private initiatives to draw on, any improvement in their lives has to be largely as a result of their own efforts (Altmann, 1997). However, many women expressed a lack of motivation over involvement in local initiatives. None of the women reported visiting neighbours or friends spontaneously today, nor inviting people into their homes, or getting together to visit events in the region. Many key informants, such as those involved in women's organizations or regional politics, believed that a considerable number of women have become increasingly socially, economically and politically inaccessible. In addition, an erosion of values and social contacts has begun to take place in their environment. Despite the prioritization of the values of solidarity, support and communal togetherness in the GDR, citizens were never 'taught' to display such attributes outside the prepared context of their working collective and, thus, never needed to maintain these values as independent actors. To some degree, women expected that *only* through employment would they experience communication, integration, acknowledgement and self-esteem, and *only* if people were employed would inequalities and jealousy in the village disappear. Hence, women expected that only another social context, equivalent to that experienced in the LPG, could resolve social problems. Women's 'wait-and-see' attitude implied a dependence on external impulses, whilst women denied themselves the opportunity to make changes that they can identify with themselves. They have not yet recognized that work does have a significant impact on the development of personal initiative but it is also linked to various other factors.

The above experiences and attitudes conveyed by women pose significant problems to their mobilization for local politics. In general, women's perception of the new democratic system was coloured by their experiences of social difference and their perception of inequality. In addition, they compared their current situation with their lives in the GDR in which they had experienced a significant level of social security and had been allocated a prominent role in public life. When they criticized democracy they raised issues mostly related to the lack of material wealth and gender issues. They felt, for example, that the lack of income prevented them from taking advantages of newly available consumer products and travel opportunities. Freedom of choice and freedom of travel remained inaccessible to them. They contended that agricultural restructuring and the restitution of property had rendered many people without a home. For these women this signified a lack of choice in their living environment. Furthermore, women did not think that people could speak their minds more freely now than under the GDR. They recalled experiences whereby women were neglected for promotion or lost their jobs because they had been openly critical about their work or employer. A significant matter of concern was the loss of control over one's own body and family planning. Many women explained that motherhood had become subject to conservative West German laws and that women were prevented from wanting children due to discrimination against (single) mothers in the labour market. Many women, with the exception of those with high qualifications, experience (and employment), concluded that the new state had achieved little for them and felt no inclination to become involved politically.

As privatization caused many workplaces, but also local shops, kindergartens and other spaces of communication, to disintegrate, there were few informal communication spaces left. Largely due to their unemployment, many women withdrew to the sphere of their private homes. The extent of this withdrawal and the mobilization of already excluded women could, at least in part, be influenced by community leaders and their capability of drawing women back into the community. Some key informants observed that female mayors were often more persistent in encouraging women to participate more in public life and generally more supportive of women's issues (Antolini, 1984). These mayors were usually women who had been 'moving spirits' in the GDR (see van Hoven, 2002) and knew how to encourage women to join in communal events. On many occasions they tried to recreate festivities similar to those in the LPG. But while local activities were welcomed as an opportunity to leave the confines of the home and reconnect with former colleagues, the active group of women consisted largely of pensioners. The activities rarely addressed the interests of younger women. When they did, women found it difficult to organize childcare and find a means of transportation. A key barrier was also their feeling of shame over being unemployed. Although many new groups and organizations on a variety of subjects developed throughout East Germany, they were not always well received. One criticism was that many groups were organized as organizations with members and membership fees. Women felt that this 'Western practice' was exclusionary. They did not usually become paying members.

Key informants regretted that a significant group of women remained out of reach, particularly as they felt that informal gatherings were essential for rebuilding social networks, improving communication and practising democratic skills. With regards to formal politics, the level of participation of women was even lower. At the local level, many initiatives were voluntary and poorly funded or not funded at all. Many women found it unacceptable to invest their own work in such work. In particular, those who were recipients of unemployment benefit or social welfare were unable to handle the associated financial burden incurred with such initiatives to cover travel expenses, membership fees and other expenses. A key complaint by women who were politically active was that they found it difficult to break through male gatekeeper structures and had few opportunities of preferment. As one woman explained, inclusion into the local council required party membership and a nomination by party members. This raised barriers, she felt, as women were actively discouraged from political participation by actors who already held positions of responsibility in existing political organizations. Women's ideas were welcomed at the informal and voluntary level but not at the formal political level.

Conclusion and Outlook

In 1989, the Wende changed the lives of women throughout the former GDR. Unification ended socialism virtually overnight and introduced a new legal system, a market-oriented economy and multi-party politics. Transition necessitated significant restructuring of socialist economic structures in order to prepare for integration into the European and global market. Large-scale privatization and rationalizations were key requirements. In particular in rural areas, where large-scale agriculture had been dominant, the human cost was high. Although men and women were made redundant to a similar extent, men were far more likely to get a new local job or to commute a distance to work. Unemployment for women soared. Unemployment became the most significant experience in the 'new lives' of rural women. In many cases, it led to increasing gender stereotyping, a withdrawal to the private sphere and an exclusion from politics at large. Many women have been unable to position themselves within the new political system whilst, at the same time, observing the disappearance of their roots in the old GDR. Rather than challenging exclusionary practices, many women complied with them. This is remarkable as this passiveness assisted in the disappearance of formerly meaningful places for women in their villages, therefore enhancing women's isolation.[4]

During my most recent visit to Mecklenburg-Westpommerania in July 2002 I found that the government had recognized that political efforts and attempts to mobilize women by voluntary organizations had not been successful. The May 1999 Amsterdam Treaty recommended that greater efforts should be aimed at promoting the equalization of men and women and the Mecklenburg-Westpommeranian Government developed a concept accordingly. In particular, this concept charged that efforts should target inequalities in the everyday lives of women with the realization that this aim was an all-inclusive task that should address all areas of policy-making, including the labour market, education, the health sector, environmental issues, culture and domestic violence. Special attention is given to the development of rural areas. The loss of employment, the

---

[4] This finding contradicts the conclusions of other studies in the rural or post-socialist context (e.g. Rueschemeyer, 1993; Scarpaci and Frazier, 1993; Smith, 1996; Graham and Regulska, 1997; Liepins, 1998; Regulska, 1998) but this can be explained by women's orientation toward their past lives rather than the future.

lack of worker skills, the need for improved transport mobility and the lack of business-oriented initiatives are more serious in these regions than elsewhere. Key aims are to improve public transport and childcare but, in fact, little is known about the actual and current needs and problems of women in these regions. The equal opportunities officer to the Government noted that structural analyses are undertaken in order to assess whether employment measures for cities could be equally successful in rural areas. Although she was positive about creating opportunities for younger women in rural areas, such as in the IT sector or in tourism, she was concerned about the situation of older women. As many of them had been unemployed for many years and had made use of few educational opportunities, she feared that it would be virtually impossible to acquaint these women with the necessary skills to survive in the labour market of today. She explained that many women needed to regain a daily rhythm, acquire communication skills and develop a competitive CV. Eventually, the equal opportunities officer admitted that it is likely that women over 50 will be a 'lost generation' for whom nothing will done.

This sentiment was also evident in my group interviews with women who had participated in my study between 1996 and 1999. Indeed, even in 2002 many women were still unemployed. Some believed they had 'simply been forgotten'. Interestingly, they had no intention of leaving their home village and seeking their fortune in the West. They felt emotionally bound to their homes in Uecker Randow but also admitted that they encouraged young people, even their own children, to leave and improve their skills and employability in the West. Whether or not their children would return, these mothers did not know.

References

Altmann, U. (1997) *Begleitende Evaluation des Pratikas-Projektes: Unveröffentlichter Entwurf zum Endbericht*, Humboldt-Universität zu Berlin, Berlin.

Antolini, D. (1984) Women in local government: An overview, in J. Flammang (ed.) *Political Women: Current Roles in State and Local Government*, Sage, London, pp.23-40.

Beer, U. and Müller, U. (1993) Coping with a new reality: Barriers and possibilities, *Cambridge Journal of Economics*, 17, pp.281-294.

Belwe, K. (1989) Sozialstruktur und gesellschaftlicher Wandel in der DDR, in W. Weidenfeld and H. Zimmermann (eds.) *Deutschland Handbuch. Eine doppelte Bilanz, 1949-1989*, Carl Hanser Verlag, München, pp.125-143.

Boje, J., Gladisch, D. and Dahms, V. (1992) *Abschlussbericht: Beschäftigungsperspektiven und arbeitsmarktpolitischer Handlungsbedarf im Arbeitsamtbezirk Neubrandenburg*, Berlin

Clemens, P. (1990) Die Kehrseite der Clara-Zetkin-Medaille: Die Betriebsfrauenausschüsse der 50er Jahre in lebensgeschichtlicher Sicht, *Feministische Studien*, 8, pp.20-33.

Conradt, D.P. (1996) *The German Polity*, Longman, Harlow.

DeSoto, H.H. and Panzig, C. (1994) Women, gender and rural development, in *Tagungsbericht 16. bis 18. Juni 1994 'Frauen in der ländlichen Entwicklung'*. Landwirtschaftlich-Gärtnerische Fakultät der Humboldtuniversität, Berlin, pp.111-129.

Eichener, V., Kleinfeld, R., Pollack, D., Schmid, J., Schubert, K. and Voelzkow, H. (1992) Determinanten der Formierung organisierter Interessen in den neuen Bundesländern, in V. Eichener, R. Kleinfeld, D. Pollack, J. Schmid, K. Schubert and H. Voelzkow (eds.) *Probleme der Einheit. Organisierte Interesen in Ostdeutschland. Band 2*, Metropolis Verlag, Marburg, pp.545-581.

Einhorn, B. (1989) Socialist emancipation: The women's movement in the German Democratic Republic, in S. Kruks, P. Rapp and M.B. Young (eds.) *Promissory Notes: Women in the Transition to Socialism*, Monthly Review Press, New York, pp.282-305.

Fink, M. and Grajewski, R. (1994) Arbeitsmarktsituation für Frauen im ländlichen Ostdeutschland, *Landbauforschung Völkenrode* 44, pp.13-25.

Graham, A. and Regulska, J. (1997) *Expanding political space for women in Poland: an analysis of three communities*, paper presented at the 'Third European Feminist Research Conference', 8-12 July 1997, Coimbra, Portugal.

Herrenknecht, A. (1995) Das Dorf in der DDR - Dorfbilder, Dorfdiskussionen, und Dorfentwicklungen von 1960-1989, *Regio*, 17, pp.4-12

Hoven, van B. (2001) Women at work - experiences and identity in rural East Germany, *Area*, 33, pp.38-46.

Hoven, van B. (2002) Experiencing democracy: Women in rural East Germany, *Social Politics*, 9(3), pp.444-470.

Hoven-Iganski, van B. (2000) *Made in the GDR: The Changing Geographies of Women in the Post-Socialist Society in Mecklenburg-Westpommerania*, Netherlands Geographical Studies 267, Utrecht.

Hudelson, R.H. (1993) *The Rise and Fall of Communism*, Westview, Boulder, Colorado.

Kolinsky, E. (1996) Women in the New Germany, in G. Smith, W.E. Paterson and S. Padgett (eds.) *Developments in German Politics 2*, Macmillan, Basingstoke, pp.267-285.

Landesregierung Mecklenburg Vorpommern (2002) *Gleichstellungskonzeption der Landesregierung Mecklenburg Vorpommern*, Schwerin.

Liepins, R. (1995) Women in agriculture: Advocates for a gendered sustainable agriculture, *Australian Geographer*, 26(2), pp.118-126.

LwAnpG (1996) *Landwirtschaftsanpassungsgesetz in der Fassung vom 20.12.1996.* Bundesgesetzblatt (BGBl.) I.S.2082

Marx-Ferree, M. (1993) The rise and fall of 'Mommy Politics': Feminism and Unification in (east) Germany, *Feminist Studies*, 19(1), pp.89-115.

McCauley, M. (1983) *The German Democratic Republic since 1945*, Macmillan, London.

Merl, S. (1991) Reprivatisierung aus der Sicht des Produktionsfaktors Arbeit, in S. Merl and E. Schinke (eds.) *Agrarwirtschaft und Agrarpolitik in der ehemaligen DDR im Umbruch*, Duncker and Humblot, Berlin, pp.109-112.

Nickel, H.M. (1990a) Gechlechtertrennung durch Arbeitsteilung Berufs- und Familienarbeit in der DDR, *Feministische Studien*, 8, pp.10-19.

Nickel, H.M. (1990b) Geschlechtersozialisation in der DDR. Oder: Zur Rekonstruktion des Patriarchats im realen Sozialismus, in G. Burkhardt (ed.) *Sozialisation im Sozialismus: Zeitschrift für Sozialforschung und Erziehungssoziologie (ZSE)*. 1. Beiheft, pp.17-32.

Pfaffenbach, C. (2002) *Die Transformation des Handelns- Erwerbsbiographien in Westpendlergemeinden Südthüringens*, Steiner Verlag, Stuttgart.

Regulska, J. (1998) The political and its meaning for women, in J. Pickles and A. Smith (eds.) *Theorising Transition: The Political Economy of Post-Communist Transformations*, Routledge, London, pp.309-329.

Reichelt, H. (1992) Die Landwirtschaft in der ehemaligen DDR: Probleme, Erkenntnisse, Entwicklungen, *Berichte über Landwirtschaft*, 70(1), pp.117-136.

Rueschemeyer, M. (1993) East Germany's new towns in transition: A grassroots view of the impact of unification, *Urban Studies*, 30, pp.495-506.

Rueschemeyer, M. (1994) Women in the politics of Eastern Germany: The dilemmas of unification, in M. Rueschemeyer (ed.) *Women in the Politics of Post-Communist Eastern Europe*, M.E. Sharpe, London, pp.87-116.

Rueschemeyer, M. and Szeleny, S. (1989) Socialist transformation and gender inequality: Women in the GDR and Hungary, in D. Childs, T.A. Baylis and M. Rueschemeyer (eds.) *East Germany in Comparative Perspective*, Routledge, London, pp.81-110.

Schumann, F. and Jahn, W. (1991) Zur Lage der Landwirtschaft in der BRD, in *Dokumentation AK Feminisierung der Gesellschaft*, PDS/Linke Liste, Bonn.

Shaffer, H.G. (1981) *Women in the Two Germanies: A Comparative Study of a Socialist and Non-Socialist Society*, Pergamon, Oxford.

Shaw, G. (1996) Women lawyers in the new federal states of Germany: Winners or losers?, *European Urban and Regional Studies*, 3, pp.257-262.

Smith, F.M. (1996) Housing tenures in transformation: Questioning geographies of ownership in Eastern Germany, *Scottish Geographical Magazine*, 12(1), pp.3-10.

Staatskanzlei (1997) *Mecklenburg-Vorpommern: Du kannst fliegen*. Schwerin.

Stark, D. (1996) Recombinant property in East European Capitalism, *American Journal of Sociology*, 101, pp.993-1027.

Statistisches Bundesamt (1997) Photocopied material obtained upon written request.

Vogeler, I. (1996) State hegemony in transforming the rural landscape of Eastern Germany: 1945-1994, *Annals of the Association of American Geographers*, 86, pp.432-458.

Watzek, H. (1997) Einflüsse der Agrarpolitik und Landwirtschaft auf die Entwicklung der Arbeits- und Lebensbedingungen (60er bis 80er Jahre), in *Leben in der DDR, Leben nach 1989- Aufarbeitung und Versöhnung*, Landtag Mecklenburg-Vorpommern, Schwerin, pp.99-119.

Wollmann, H. (1995) Regelung kommunaler Institutionen in Ostdeutschland zwischen 'exogener Pfadabhängigkeit' und endogenen Entscheidungsfaktoren, *Berliner Journal für Soziologie*, 4, pp.497-514.

Chapter 8

Can Education be a Strategy for Developing Rural Areas?

Ingunn Limstrand and Marit Stemland

Introduction

This chapter is written in a Norwegian context, and focuses especially on education as a factor for regional development in rural areas in the northern part of Norway. The authors' base is The Northern Feminist University, a small institution for knowledge, education and competence about gender, women and women's lives in the North. The Northern Feminist University is located in Steigen, a small municipality an hour's boat trip north of Bodø – very much in the peripheral North.

Traditionally a fisherman-farmer community, Steigen with its 3 000 inhabitants faces the same challenges as many small rural societies. Population and birth rates are in decline, and changes in workforce, production modes and education patterns tell of a need for economic and political transformation. Yet, through the power relations that have developed between the urban and the rural, the centre and the periphery, the northern periphery is denied the opportunity to define its own solutions for development. The production of knowledge is professionalized, commercialized and centralized, and, all in all, takes place in the centre as opposed to the periphery. A marginalization takes place, in which northern inhabitants find their everyday lives affected by events taking place somewhere else and by decisions made far from where their effects are felt.

In this context, the situation of women is especially important. Both women and men live in small-scale rural communities, but in many cases it is the men that most readily find work. Jobs in fishing, agriculture and forestry are conceived as 'male', and, in a society where higher education is available to most, women find their opportunities in rural societies limited. Illustrative of this, a recent study for Nordland County showed that, of the women from rural communities aged 19-24 who were interviewed, only one in 10 could envision moving back to her childhood municipality. Most wanted higher education and a challenging career.

They did not see the small, rural community they had came from as a place where this was possible (Nordland Fylkeskommune, 2003).[1]

Without people, rural communities cannot survive, and there have to be opportunities for both women and men if societies are to thrive. In this chapter, we are looking at the possibilities rural areas possess for providing citizens with the lives they want and how this can come about through conscious efforts on the part of planners. Our focus is on education as a strategy for development and on the role of the Northern Feminist University in Steigen municipality, as one example of innovative action. The main objectives of the chapter are:

- To introduce another perspective and strategy for *developing* rural areas, from the position of a peripheral community.
- To raise questions, and start a discussion about *education* as a strategy for rural development.
- To stress the importance of being *gender-sensitive* in developing strategies for rural development, focusing on women's needs and wishes.

Analytical Concepts and Approach

First, we need to provide a brief commentary on the perspectives or concepts that are used as a base for the discussion presented in this chapter. The first of these perspectives or concepts is that of *situated knowledge*. This is understood to be knowledge that is defined from our own personal contexts, which in the case of the authors includes our position as rural citizens. This is a knowledge that makes no claim to be a producer of universals, as positioning in itself creates responsibility for the producer of knowledge (Haraway, 1995), with individuals that are the subject matter of an inquiry recognized as acting subjects.

The temporal framework in which subjects are acting is that of *modernity*. Modernity is characterized by rapid change, urbanization, globalization, the transformation of culture, and time-space compression (Massey, 1991). Further, in the Western world, modernity is associated with the welfare state, materialism, calls for equality, searches for greater knowledge, plus demands for more information and democracy. Hence, the implications of modernity for individuals could include increased levels of mobility, where space barriers are broken down, plus the availability of means of communication, media and market, such that strong linkages go beyond local and national boundaries (Massey, 1991). Individuality and individual freedom of choice is increasing, with reflexivity

---

[1] Kvinnemeldinga: A report from Nordland Fylkeskommune (Nordland County) based on studies by the Nordland Research Institute and the Norwegian Institute for Urban and Regional Research in Finnmark. The mandate of the project was to raise knowledge about women and gender equality in the northern part of Norway, focusing on analyzing the situation in general, to give an overview of affirmative action for women in the county, and to evaluate existing and new actions in the county. This report is to form the basis for further planning in the county.

regarded by some as a defining characteristic of all human action in the modern world (Giddens, 1990; Limstrand, 1996).

One element in this reflexivity is a greater understanding of *gender*. In this regard, the development of different strategies and affirmative actions in rural development are connected to three different theoretical approaches to gender (Pettersen, 1999). These three approaches are:

- **The Equal Opportunities Perspective**: In this theoretical approach, women and men are seen as equal in principle, as human beings. Differences between women and men and their differential treatment are seen to be discriminatory; to be undesirable in an equal society. Within this framework, gender equality is a question of human rights. By granting women the same legal rights as men, the existing inequity between the sexes is expected to disappear gradually. This approach is the basis for official gender policy in countries like Norway. From this perspective, strategies for change are connected to making women and men equal by giving them the same rights and opportunities. In this interpretation, gender is a variable or an indicator (Pettersen, 1999).[2]

- **The Women's Perspective**: This approach builds on the idea that women and men are different. They are seen to embody different values, attitudes and ways of thinking, with differences between women and men seen as a critical potential for change in society.[3] The concept of women's and men's different rationalities (Skjønsberg, 1995), studies of the gendered division of labour (Ellingsæter, 1995; Ellingsæter and Solheim 2002), investigations of men's dominance and women's subordination in private relations and in society's institutions (Lundgren, 1985), are some examples of research that is based on this perspective. The idea at the heart of this perspective is that highlighting and valuing women's experiences and values should create equity and equality between women and men (Pettersen, 1999).

- **The Gender Perspective**: In this understanding, men and women are in principle neither similar nor different from each other as groups. Women are different from each other, and should not be considered a homogenous group, with the same applying for men. From this viewpoint, the understanding of

---

[2]   The Gender Equality Act of Norway (1978) has a gender-neutral principal rule, which is that difference in the treatment of men and women is illegal. However, differential treatment may be permitted if that treatment promotes gender equality. Hence, affirmative action in favour of women is a measure to improve existing inequity between the sexes. Such action is not undertaken to give preferential treatment to certain groups. The use of affirmative action in favour of women has been chosen as a strategy because it was recognized that it is not possible to achieve gender equality merely by prohibiting discriminatory treatment.

[3]   An important critique of this perspective is that the assumption that women and men have different values, attitudes and ways of thinking could lead to essentialism in how we view women and men. Gender becomes determined and there is little room for individualism.

what it entails to be a woman, or a man, is continuously changing, for gender as a concept is not definite; its content is subject to change. Hence, we need to deconstruct and reconstruct, to breakdown and redefine gender categories, looking for understanding and definitions of gender (or gender identities) that are connected to different contexts (Pettersen, 1999).

Working from a gender perspective gives us the opportunity to develop a tool for understanding gender as an integrated part of processes and structures that were previously considered gender neutral (Harding, 1991; Widerberg, 1992). The main objective is to empower women and give them real opportunities, flexibility and options. Gendered divisions of labour, gendered structures and strongly gendered attitudes and identities need to be analyzed, redefined and changed. This is, we feel, best approached using a *relational* perspective. Here, gender is continually produced and reproduced in relations between actors and in broader social relations within society. The meaning of gender is renegotiated in each individual case, which may involve negotiations between women and men, between women and gendered attitudes and identities, between women and gendered structures, etc.

Power

In this process, power is a central dimension. Power is a relational concept, in that power is not something people possess but is executed in social relations, in which strategy and struggle is employed to see interests achieved. This perspective gives room to see people's own will and force as elements in social change. As for the position of gender, here a relational perspective on power does not necessarily mean that options and opportunities for exercising power are evenly distributed amongst women and men. Not only are opportunities for power exertion unevenly distributed, for so is the will to engage in such action. Options for executing power are often linked to position, but the will to execute power is linked to an individual actor's willingness to pay the price of doing so (Lotherington, 2002).

Both positions of power and the cost of executing power are unevenly distributed. Men as individuals may not *possess* more power than women, but as a group they have more opportunities to *execute* power. This arises because men more often hold powerful positions, and gender structures that are prevalent in society favour men over women, with the result that it is less costly for men to execute power. When a man acts in a forceful manner, he is commonly perceived as acting positively, by being a man. A woman who does the same – even when she holds the same position – is likely to be conceived negatively, by acting in an unfeminine way. The construction of masculinity as more powerful than femininity favours men as a group, and may make it more difficult for strong women to exercise power in positions they hold (Lotherington, 2002).

This power to make conditions in society seem natural and obvious is the power of discourse.[4] Politically produced, discursive power influences how

---

[4] Foucault (1999) ties discourse to power and knowledge. Power is obtained by those who decide what and how things are to be discussed. In this chapter the concept of discourse

individuals perceive themselves in the world, including our perception of who should be defining strategies for development. In the prevailing discourse, this power has been 'given' to urban areas, leaving rural districts as recipients of policies that may or may not suit the 'realities' that exist away from big cities (Eidheim, 1993; Rossvær, 1994; Ellingsæter and Solheim, 2002; Makt-og demokratiutredningen, 2003). It is a general tendency, both nationally and internationally, that the centre in increasing degrees defines the periphery, economically, culturally and politically, leaving the rural areas with less possibility to mobilize social and political resources to assert themselves and to develop (Aasjord, 2003).

District Policy and Education

In this section, we take a closer look at the Norwegian district policy for education, and discuss this in relation to rurality, modernity and gender. Our point of departure was introduced at the beginning of this chapter, which is that we live in a small community, working at an institution that is primarily concerned with capacity-building and education for and about women. As an appropriate backdrop to this, it is worth citing part of the Norwegian Government's White Paper 34 on local and regional development in Norway:

> While access to jobs is still the greatest influence of where people choose to live, these choices are more than before driven by the sum total of services offered, leisure time options, accommodation and jobs. A lot of businesses in more rural areas have problems recruiting competent workers, while at the same time there is a scarcity of jobs for highly educated people. (Kommunal- og Regional Departmentet, 2001, chap. 1.2. our translation)

The basic idea of the Norwegian district policy is to maintain a dispersed population. Traditionally, rural areas have been characterized by being far from decision-making areas (e.g. the capital and bigger towns), by a sparse population distribution, a dependency on primary production, like fisheries and agriculture, by a lack of workplaces for people with higher education and a lack of skilled labour, as well as being socially different from urban areas (Almås, 1995). Peripheral communities lag behind in development terms because, to an increasing extent, activities like business, education, services and health-care are located in and around bigger towns and regional centres.

A case point is the increased number of people who live in centres with higher education and high-income jobs compared to rural areas (Sørlie, 2000). This development increases the social, cultural and political marginalization of the Norwegian periphery, and through the terms of discourse today the centre gets to define the terms of debate. By and large, strategies for rural development have

---

is given a broad meaning, including written texts, communication and social practice produced by human experience and action.

traditionally focused on (low-level) blue-collar work and work that does not demand higher levels of education.

But modernity has introduced better telecommunications and infrastructure, a wider geographical distribution of social relations and easier access to the media. On the whole, this has changed conditions in rural areas. Improved communication means that people can stay in touch, that they can be connected independent of distance and location. In addition to which, dependency on primary resources in rural areas has lessened. In this situation, it is important that the government defines a special policy and strategy for developing rural areas that lie outside urban regions. This policy should be sensitive to the needs of a *modern* population in peripheral communities. In order to achieve such development, the process of rural and regional development must change towards being more inclusive and encourage bottom-up participation. In a modern society, all policies, as well as mainstream policy actions, need to be analyzed so we can understand and highlight their consequences for the rural population (much as the Countryside Agency promotes through 'rural proofing' measures in England; e.g. Countryside Agency, 2002).

Turning to district policy and gender, there are three points to make that describe the structural situation. Firstly, there is a difference between women's and men's educational levels. Up to secondary school-level there is no difference between Nordland and the rest of Norway. But at graduate level, whereas 16 per cent of the population aged 16 years or more in Nordland County have higher education, 21.3 per cent have the same in Norway as a whole. Gender-differentiated, higher education has been achieved by 16.6 per cent of women and 15.3 per cent of men in Nordland, while the percentage is 21.3 per cent for both women and men in the country in general (Nordland Fylkeskommune, 2003, p.16).

Secondly, women's and men's choices in education and working life are to a large degree divided along gendered lines. On the whole, women are employed in the public sector, in caring professions, in administration and teaching (as in Sweden, see Scholten, 2004). In the private sector, women dominate the workforce in the commodity and hospitality trade.

Thirdly, almost half the female labour force works part-time, with 45.2 per cent holding part-time posts in 1998 (Vikan, 2000). This tendency is more pronounced in rural areas, with only 56.2 per cent of women in paid-work working for more than 30 hours a week in Nordland, compared to 85.6 per cent for working men. For some municipalities the number of women who work full-time (>30 h/week) is as low as 36.4 per cent (Vikan, 2000, p.17). In addition to this, women still carry the main responsibility for domestic work.

The average income of working women is lower, whether wages are seen as gross or net earnings, or as a percentage of men's earnings (Vikan, 2000, p.18). As shown in the Nordland County 'Kvinnemelding' referred to above (Vikan, 2000, p.20), the county has many small municipalities with few women in the 20-39 age group. As one illustration, in Moskenes there are only 65 women per 100 men in this age group (Vikan, 2000, p.15). Although more women than men in Nordland now have higher education, education levels in the county are still below the national average. Women in Nordland work less than average, and more work part-time, and, possibly as a result of this, earn significantly less than men. Added to which, there are fewer women in leadership positions, as seen amongst board members in industry, etc., while in politics Nordland has the same gendered representation as the national average (Vikan, 2000, p.19). This structural picture creates and maintains barriers for women's development.

Analyzing the population in terms of gendered divisions gives a clear structural picture of a rural society with problems. Men and women in the demographically very important 20-39 age group tend to leave for towns and educational centres, and many do not return. In seeking to address this issue, political strategies and rural development actions have focused on women, but have these efforts been effective?

Strategies for women and rural development can be understood in the light of the three perspectives on gender presented earlier. In the 1980s and early 1990s, the *women's perspective* dominated ideas in planning and development thinking. The practical strategy that was promoted in association with this perspective was to use affirmative action to give women the same opportunities as men – a mixture of the women's and equal opportunity perspectives. The focus at this time was to develop strategies and models for making women want to settle in rural areas, mainly by offering training for entrepreneurship and job-creation, rather than training that would directly increase their technical skills and formal competences (Alsos *et al.*, 1999). Lotherington's (2002) critique of this strategy is that women were used to fulfil a predefined strategy, and were not given real influence in defining the content of policy or development.

From the late 1990s, ideals of gender mainstreaming and integration have dominated the picture. Strategies today are based more on a *gender perspective*, and, to a certain extent, take account of the need for more formal capacity building, for decentralized education systems and for flexible models based on the needs of women in rural districts. However, as this strategy has been difficult to implement in rural areas, these mainstreaming models have helped maintain the development imbalance between centre and periphery. As a consequence, rural development policy remains an 'in-spite-of' strategy to reduce the consequences of mainstreaming actions. In sum, this means that modern society is creating new conditions for the periphery – as seen in (power) relations between centre and periphery – for women living in the peripheral rural areas, for education and education patterns and for women's choice of education and work.

Education as a Strategy

Our opinion and experience is that a focus on education must include a strategy for rural development; that any strategy on developing parts of everyday life in rural areas must include and be seen within a rural development perspective. In arguing this case, we begin by discussing development patterns and attitudes towards education and capacity building in rural areas. We examine these in the light of modernity and changing conditions in rural areas, as well as in terms of discursive power relations. Gender and gender roles remain an underlying variable throughout this commentary. We also discuss competence-raising in connection with primary production. Finally, we use the Northern Feminist University (NFU) and some education models developed in a rural health project carried out at NFU as best practice-models, to illustrate some new ways of thinking about rural development and education.

Attitudes Toward Education and Competence

Traditionally, settlement in the rural areas has been located where natural resources were available. In coastal areas like Steigen, fisheries and small-scale farming formed the basis for settlement. Exploitation of natural resources and, to a certain extent, the subsistence economy did not require high competence, in terms of an academic or formal competence. But in more recent generations, an increasing number of young people – many from rural areas – have sought higher education. However, rural areas have not had a need for skilled labour, which has led to a situation where rural areas find themselves exporting youth. Young people who choose not to pursue higher education remain in the local community, while those who leave for higher education courses rarely return. This has created a view of 'educating oneself away from the rural areas' – or that *education is a threat* to rural areas. In this perspective, knowledge and education become a threat to the periphery, in that a long-term effect of higher education is the depletion of youth from rural societies.

As for the gender situation regarding jobs in rural areas, we find that the professions in resource-based businesses have to a large extent become *male professions*, while women's work is made 'invisible' or given lower status. A typical example is the fisheries. In traditional households in North Norway, the husband was away fishing for extended periods of the year. During these periods responsibility for the children, home, livestock and harvest fell on the wife, who was also responsible for getting most of the equipment ready so the husband could go to sea in the first place. Her efforts were as much a part of the family economy as his, and no less valuable for family subsistence. From a historical and cultural perspective the fisheries sector has always been extremely gendered in its organization. This, together with the gendered organizing of farming, may have influenced the Norwegian labour market in general, making Norway one of the most gendered labour markets in Europe (Birkelund and Petersen, 2003; Likestillingsbarometeret, 2003). In spite of a cultural change in the 1970s and 1980s (geographical and regional differences), this traditional way of organizing households still exists in fishing villages like Steigen (Otterstad and Jentoft, 1994; Pettersen, 1994). This creates and upholds traditional gender roles. It would be interesting to look closer into the tension in a life between traditional and modern household.

In a modern fisheries family, the husband is still typically in charge of fishing. However, in addition to responsibility for the home and children, a wife is now likely to hold a wage-paying job that supplies cash income and security for boat loans and a house mortgage. If she works in the fisheries industry, it is most likely in an administrative position or in a fish processing plant. Many women also keep boat financial accounts and do associated paperwork. Even so, her efforts are not recognized as part of the 'fisheries', unless she works in a fish processing plant. Much of the same is true for agriculture and forestry, other mainstays of rural communities (Blekesaune *et al.*, 1993).

Today, the jobs that local rural communities can offer young women are mainly in health care services, in schools and in public administration. But many young women have other educational plans and wishes for their lives, as shown by the fact that the municipalities in Nordland that have more young women than men are those in which high schools and university colleges are located. In effect, the concept of educating oneself away from rural areas is truer for young women than for young men. Indeed, the power of this dominant discourse means that it becomes more difficult for young people with high education levels to see ways of moving back to their home communities (Hansen, 1998). While this perspective is allowed to hold ground, the periphery will continue to 'export' young people who should be key resources in their development.

All things considered, if rural areas are to be sustainable societies, their economic, social and political situation needs readjustment and development. But readjustment in itself is competence-intensive, simply because it demands knowledge of, for example, leadership, organization, development of resources and economy. With higher levels of education, the chances of both society and individuals being more flexible increases. With reference to the economy, flexibility means people are more able to adjust to change, for example in the market, and that they are also in themselves more able to influence the direction in which any adjustment turns. As regards democracy, it is claimed that increasing the knowledge, education and discursive capacity of participants means that they are better able to take part in development (Kylhammar, 2003; Makt- og demokratiutredningen, 2003).[5]

As regards the social situation, for the periphery to develop a post-modern society, its citizens must be given space for diversity. This goes for diversity in lifestyle as well as in gender roles. In such a society, higher education must be seen not as a threat, but as an opportunity.

Finally, the political status of rural areas may be better placed with higher education levels. This is based on the fact that national and regional social planning is conducted by a highly educated administration. To influence this process, the periphery needs competence to be able to play by the same rules, and be

---

[5]   A Parliament report on power and democracy in Norway was completed in 2003 after five years of research. Several hundred researchers were involved in the work, with 50 books and 77 reports produced. The main conclusion from the report is that all links and levels of representative democracy have been reduced. At the same time the legal system increases its power. The most important change in power relations in Norway is that democracy as representative government – a formal decision making system that works through majority votes and elected bodies – is decreasing. The political spending power of the voting slip is reduced (http://www.maktutredningen.no).

considered entitled to give an opinion. Briefly, before turning to our main focus on the potential for developing new sectors, we wish to raise questions concerning competence-building in traditionally-based production businesses in the primary sector.

The Traditional Economy: Primary-Based Industry and Activities

There is a need for documenting, valuing and upgrading the competence that has been developed in the resource-based economy through the generations. In marketing products, it has become necessary to be able to document both the production process and the competence of the labour force. The more complex system for distribution of fish for international markets, demands and regulations for processing of fish and regulations for business development and organization push forward a more theoretical and formal knowledge. In the fishing industry and in fish farming/aquaculture, for instance, many experienced workers take education and get this theoretically based education. Their informal experience combine into formal, documented competence. This is valuable for employees as well as for firms.

In the agricultural sector, education in the form of agricultural college courses has become more common. The authorities push forward the need for formal competence in entrepreneurial, technical, economic topics, and these topics are offered in short-time courses. But due to economic needs, more and more farmers turn from full-time to part-time work in agriculture. In such a situation, many will give priority to educating themselves into a wage-paying job, and take the farm as a side occupation. This is not a good starting point for developing agricultural businesses.

Another strategy that can be used to increase the production of value in rural areas is the further processing of raw materials. If we do not succeed in this, the population can be expected to decrease owing to a fall in employment in primary production itself. Technological development often leads to a reduction in the number of employees needed just to deliver raw materials. Yet, to develop the marine sector, competence-building is essential, because further processing and technology development demands skilled labour. If the fisheries sector is to become attractive to young people it has to be based on theoretical knowledge and education, and demand for higher education and theoretical competence, with various jobs related to these needing to be given much more focus when marketing the fisheries to young men and women (Thordarson, 2003). Due to the globalization of the economy, we can assume that the price of agricultural commodities will fall. To keep up farmers' incomes, and maintain employment in this sector, there is a need for further local processing and marketing of farm products. This requires a different form of competence in the agricultural population, and education is required to make this possible. Later we will come back to education strategies that are relevant for rural areas.

Higher levels of education in general and in agriculture in particular, along with increasing levels of processing in local (small-scale) farm units, offers the possibility of new employment in rural societies. Women wanting to combine farm household life with starting or running their own business may find opportunities here.

New Sectors: Developing Competence Based Industry

In terms of possibilities from the existing economy, there is a need for research in order to develop new products and new production processes. Very little research and development actually takes place in rural areas, because most relevant institutes are located in cities. Even though some are located in northern cities like Bodø and Tromsø, there is a great difference and distance between the rural north and rural northern cities (Aasjord, 2003). Seen from a rural point of view, most research about Norwegian society is carried out from established academic milieus in central cities, which are far from everyday rural lives. It is essential that more of this knowledge production takes place in the periphery; so creating closer interaction with the people, businesses and lives rural inhabitants are living today. In the social sciences, the researcher's context should not be underestimated. Academia and the big cities are not necessarily the best context for every topic of social studies. In local-community research and rural research by social scientists in particular, there is a danger of researchers being lost in conceptions and ideas about the periphery that are grounded in how it was 10 or 20 years ago, perhaps when researchers lived there or left it.

By locating research in areas where values are actually produced, we see a potentially fruitful interaction between research and production. Such interactions can influence which fields research focuses on, as well as the results and spin-offs from such research. To a larger degree, this strategy will ensure that research results reach out and are utilized in the periphery. Here we are talking about applied research, which identifies need for change, development and further research, in addition to describing the current status of rural areas. This means that the bonds between academia and the locale remain strong, and the conditions for doing research improve as its usefulness becomes visible.

We can also draw a parallel to the political sphere and the important interaction between politics and knowledge production, for *knowledge is strongly connected to the power of defining the agenda*. Those who have the formal competence and who develop canonical knowledge are those who set the agenda. In this perspective it is essential for the periphery to take part in knowledge production. If rural societies continue to see knowledge, education and knowledge-production as a threat, a vicious circle of dependency on central areas, and increasing lack of power to set the agenda, could develop.

In terms of discursive power – the power of what is so self-evident that it determines not only how we see issues but also what we see in the first place, thinking about the peripheral north works to mould the region into a preconceived shape. By defining rural areas as traditional-minded, dependent on labour-intensive resource-based primary production, that is how reality is shaped. Thinking that there is no future for them there, young people with higher education do not seek to establish a future outside urban areas, and the dreaded 'brain drain' maintains the image of rural areas as zones of low-competence. In sum, the prevalent discourse of the rural as marginalized serves only to marginalize these areas further. This leads to apathy, defeatism and a further depletion of vulnerable societies.

Our contention is that, rather than accepting this situation, new strategies must be developed to ensure that rural societies can partake of the changes brought about by modernism, and that people in rural areas can live the lives they desire with a full range of options for themselves. One way of achieving this is to break routines by changing behaviour patterns. As we show below, developing competence-based businesses can be a successful strategy for rural areas, so we turn to a concrete example, the establishment and activities of the Northern Feminist University in Steigen, to illustrate this point. We do this not simply in order to situate ourselves as researchers and practitioners, but also to show that this institution in many ways represents an attempt to introduce new concepts and initiatives in the peripheral north.

The Northern Feminist University: Education in Rural Development

Established in 1991, the Northern Feminist University (NFU) is a private foundation. Situated in Steigen in Nordland County, between Bodø and Narvik – remote from and peripheral to almost everywhere, the NFU is not a traditional university as such. Rather it is a centre of knowledge, which focuses on gathering and collecting knowledge of women, by women, for women, with an emphasis on upgrading this knowledge, developing it, documenting it and teaching it. The NFU works with researchers, grass-roots women, politicians, planners, etc., in seeking to disseminate knowledge to those environments it concerns. The intention is to be a meeting-place between formal and informal, theoretical and practical knowledge, and between women from all levels of education and society. The NFU works both locally, regionally, nationally and internationally, in that we do not live in a vacuum, and that what is done, changed or developed in some societies will infect and influence the situation of women in other parts of the world.

The NFU was established in the context of the situation in rural North Norway in the 1980s, where:

• The level of formal education of women was lower than that of men, and was lowest in rural districts. The problem of centralization, with higher education offered in urban centres only, existed then as now. Indeed, research then was limited to universities and colleges in urban areas, even if the subject matter was concerned with rural topics and problems. In 1986 figures show that 55–60% of adult women had only compulsory education, and that fewer women than men in North Norway had education at a higher level (Woie Berg *et al.*, 1993).

- Young people leaving rural areas, and women in particular, were moving to urban areas to get an education and work, leaving older people and men in rural districts.
- Rural women's choice of training was very traditional, mostly within the health and care-related professions, in teaching, in service professions and in lower-level administrative work.
- Women's attitudes towards the combination of education, working life and care tasks was to see care of children and the family as a higher priority than education and paid-work.
- Attitudes toward education in rural areas were grounded in a belief that education took people away from rural areas, and that rural areas had no need for higher education and theoretical competence.
- Few women were leaders at higher levels within organizations or institutions, whether in the private or the public sectors.

At the end of the 1980s, the state and the government realized the problems these 'womenless' men and 'womenless' local communities presented, and decided to take affirmative action to encourage women either to stay or to move back; more or less communicating the view that women were a kind of infrastructure that needed to be invested in for the benefit of men or the district. At this stage, women were seen as the problem, because they were not there to care for men and did not provide services that were needed in local communities (Pettersen, 1994).

These strategies were developed from the aforementioned *women's perspective*, seeing women as different from men and therefore a useful contribution to the community. Slowly, the picture evolved, from that of women as a problem (by leaving) to that of women as a resource. (Now the paradigm turns again, with young, well-educated women looking for jobs in such areas, as places where they wish to settle, so rural districts should now regard themselves as resources for women.)

Nordland County was very keen on these affirmative actions, and started a lot of women-directed projects in the 1980s. They lasted for some years, then the projects finished and the project leaders left to start other jobs. By 1990 all the experience, knowledge and competence that had been developed was floating in the air, difficult to utilize for new projects or in 'real life'. So the County needed an institution to take this knowledge, use it and disseminate it. Given the education gap between men and women, alongside women's commitment to family and care tasks, and barriers for women in rural areas to attend higher education, it was decided that what was needed was a specialized institution to work on these issues.

The economic situation of rural municipalities was as problematic in the 1980s as it is today. Steigen municipality was almost bankrupt at that time, and was 'screaming' for more money from the state. The state refused to sanction further funds, but did offer a project on how the community could manage on its own.

Through that project it was found that the community had three resources on which to build self-management: namely, *youth*, *identity* and *women*. The ideologist of the feminist university, Berit Ås, was invited to tell the community about a newly started feminist university in the south of Norway, and a support group for a feminist university was established in Steigen. 'You're crazy!', some people said. But even though they did not quite grasp the idea, the political and administrative leaders of the municipality provided support and helped with approaches to the County administration. Both municipality and County believed in the idea of having a feminist university in the 'real periphery', and, together with other founders, gave money for the start-up. Established in 1991, the Northern Feminist University is not a big institution. There are only 12 people working there, and only five work on gender issues. Funding comes partly from the national government budget, partly from projects, studies and courses, and partly from a guesthouse where course participants stay, that operates as an ordinary guesthouse when there are no or few students. The NFU undertakes project work, documentation and specific tasks for ministries, counties and municipalities, as well as providing part-time courses at all levels. As the NFU has no examination rights, there is always cooperation with a high school, college or university on formal studies. These are open to men and women, with several male students on part-time courses, although so far none on gender studies. Special women's courses are open only for women.

Through its work, the NFU has based its activity on the three gender perspectives (Pettersen, 1999), adjusted its activity to the adult everyday life of women in rural areas, and used adult pedagogies, counselling and communication methods as working tools. It has focused on themes such as social planning, rural women, resource management, the need for competence in rural areas, violence and sexual abuse against women and children, and through this achieved documentation and project competence. The NFU has offered education at a high school level to a lot of women in the local community who would otherwise have had to leave to get similar education. Several of these women have later taken higher education courses within a decentralised model, and have got jobs in the municipality.

The NFU also offers higher education within the fields of leadership, social planning, teachers' training, nurses' training and languages. And it has initiated and run projects that have given working possibilities to women with high competence. It has carried through both regional, national and international conferences and seminars, and contributed in seminars and conferences in Europe as well as in Africa. With partners from EU countries, the NFU produced a video, *Northern Women – New Images*, about women with high competence who have chosen to settle and create challenging jobs in rural areas. The video is distributed to high schools and other environments in the four partner countries.

The NFU has developed into an equal opportunity competence centre in the region, which the County Council has suggested should become the public equal opportunity centre of the region. In addition to all these things, the NFU has developed flexible education models and given value to the concept "rural competence".

Through its work we feel that the NFU contributes to increasing the region's consciousness concerning equal opportunities and policies for rural women. We do not think it is accidental that Steigen this year got its first female mayor, and has a local council with a majority of women. Five of the eleven highest ranking municipalities with a majority of female counsellors come from the county of Nordland (Likestillingsbarometeret, 2003).

The main NFU focuses are on democracy and social planning from a gender perspective. Networking with women's projects and organising occurs, nationally as well as internationally – in the Barents Region, elsewhere in Europe, in Africa, in the circumpolar areas and in India. Most work is, however, community-directed, which is the arena we now turn to in order to explore some projects in detail.

Strategies for Education in Rural Areas

In the final part of this chapter, we introduce some strategies for providing education in rural areas along cost-effective lines, while maintaining a focus on education for communities and their citizens, so as to ensure fulfilled lives in healthy, sustainable communities. These strategies include acknowledging rural experiences and contexts, both as regards taking into account their limitations and advantages. One way in which this rural sensitivity is expressed is in the running of 'combined classes'. These have emerged from the recognition that a declining rural population results in a smaller allocation of economic resources to municipalities, so they have fewer services, including education. As a result, schools cannot afford to establish classes in auxiliary work, in care work, for ambulance staff or for any other group for which there are few students, as average costs per student are too high. Yet, municipalities have no need for a full class of students in every profession, because they cannot offer all students a job. But municipalities do need representatives from different professions and with a variety of competences. Recognizing this conundrum, the NFU offers combined classes that include related professional training in one class, giving students shared tutoring on topics that are common for professions, with auxiliary tutoring for smaller groups on common topics and separate tutoring for those topics that cannot be combined. These combined classes can take place in different municipalities, with common tutoring occurring through video-conferences from the same 'mother school'. A greater part of the education is carried out as self-study, or in practice within a workplace with a counsellor. By offering such courses, an important focus for the NFU has been to develop capacity in society, so as to build competence in the population that can contribute to local municipalities in general, while giving people opportunities for fulfilling lives. One example of this in action has been the 'Helse i Utkant' (Rural Health) programme, where training auxiliary nurses, nurses and other health care workers for the needs of rural society was central. This was part of the overall capacity-building plan in Steigen.

Another example from the health service field is a project to develop health professions and health care work in rural areas into more challenging and interesting work, at the same time making it more effective and improving quality. Typically, rural areas are in constant need of doctors, nurses and other health professionals, as there is a high turnover in these positions. Many leave because the professional environment is limited, with some people leaving because they feel lonely (professionally) and want a professional network to discuss everyday challenges. Together with the health care institutions in Steigen, the NFU has developed an interdisciplinary counselling programme (Counselling as part of your job) that has become a forum for both reflecting on one's own practice and routines, as well as professional discussions between different professions on various professional topics (Sivertsen *et al.*, 2003). Technological communications solutions have also been employed so that smaller communities can work together to create a network for health care workers. Our counselling as part of your job project could be adapted to work with technological solutions.

These professionals develop an extensive competence because they are far away from specialized institutions and have to deal with more problems on the spot than do health care workers in cities. One of the strategies in the NFU's health care projects is to engage rural doctors as lecturers for medical students at university, both to upgrade these doctors' rural competence and give students a more realistic picture of the interesting and challenging work in rural areas, as well as providing some preparation for work in those areas. It is hoped this may make rural doctors value their own competence and range of experience more, thus resulting in capacity-building and a strengthened identity as a 'country doc'.

Concluding Remarks

Our main objective at the NFU is *to develop strategies for change*. This has always been the main focus of NFU research about gender and women. The development of knowledge has been closely connected to the development of policies and strategies for change. The aim of researchers has been to produce knowledge in order to empower women, and also to empower the northern periphery and redefine the discursive power of the centre.

Primarily, we need a holistic, situated perspective. To be able to change the system and its structures, and in order to develop rural areas, we need to analyze the situation from the perspective of these areas. The political climate and central tendencies in state rural policy today are not envisioned from the perspective of these areas. Rural policy has become an 'in-spite-of' strategy, aimed principally at reducing the consequences of development. There is an important need for capacity-building, both in traditional production spheres and in potential new sectors. Situated knowledge is a presupposition for reconsidering the power of definition (discursive power), in the social, cultural and political development of the periphery. But rural societies in general must realize their need for educated and competent people, and take charge of their own development process.

The development of telecommunication and infrastructure means that distance is no longer as big an obstacle for locating knowledge-intensive businesses in rural areas. Advances in communications technology and inputs from a world that is rapidly changing push forward the need for developing rural areas, but we must not forget the importance of creating space and options for diversity. New conditions and modernity necessitate an alternative development, a development that creates space for multiplicity within peripheral communities and gives inhabitants opportunities to live a modern life. The picture today shows that quite a few young and highly educated people look to the periphery for a place to settle and develop what they see as 'a good life' The upcoming generation seeks a diversity of job opportunities, a varied cultural life and a tolerance of different attitudes. Should they choose to settle in a small community, they still demand the right and possibility to be citizens of the world. This is a great challenge to a small-scale community, which traditionally has had the image of conformity, in the meaning that most citizens had the same background and worked in similar trades. From our point of view, it is crucial to find a strategy to develop modern and inclusive societies in the periphery.

Norwegian regional development policy highlights the fact that a gender perspective is a condition for developing viable communities (Kommunal- og Regional Departementet, 2001). We need gender knowledge and analysis to develop a strategy for rural development, meaning gender-disaggregated data and gendered analyses of plans, decision-making bodies and gender distribution in general. Gender differences underline the importance of being gender-sensitive when developing strategies for education and rural development. One example of which is the public sector, where the regional division of workplaces is of crucial importance for women's residential choices. Thus, if jobs were available in rural areas, highly educated women may choose to settle in rural areas to a greater extent than men with comparable levels of education, because many women work in the public sector.

Looking at Steigen, attitudes in general and gender roles in particular are still relatively traditional. This may lead to presumptions that women are responsible for caring tasks in society and are not expected to hold decision-making positions of power. These gender identities have implications for and to a certain extent define women's choice of work and careers.

Educational levels in Norway are rising, with women a big part of the educated workforce. To attract women and increase formal competence in rural areas, it is important to change attitudes and thinking about education, competence and rural areas. One solution could be decentralized education models that meet individual women's needs – like women who are single parents and need to work but also seek education to empower them. The dominant rural point of view today focuses on resource-based industries, and, as mentioned above, this work is very male-dominated. Yet, even here, there are job opportunities in rural areas that are comparatively high-income (fish farming in particular). Making space both for women and men in such occupations is a way to make industry and communities viable in the long-term.

We need to analyze the situation of women from the perspective of the negotiation of women's gendered identities at different levels, as these shape their lives and opportunities. Women's needs must be taken into consideration. This is a question of democracy, and of justice. Women's skills and competence must be recognized as a resource. A strategy for development must highlight the understanding of different preferences and needs according to gender and age. We need to focus on state policy in particular, both for rural development in general and for the treatment of women's issues, legal rights and the machinery for promoting these on different levels.

As a final conclusion we might say that the establishment of NFU has been a successful strategy to highlight the need for educated people in the periphery. NFU has contributed to the development of good models for a sustainable development in rural areas based on competence-building and education, and has become a centre of activity in what remains a lively small-scale rural community.

References

Aasjord, B. (2003) Hvem definerer Nord-Norge? *Dagbladet*, 13 February.

Almås, Reidar (1995) *Bygdeutvikling*, Samlaget, Oslo.

Alsos, G., Anvik, C.H., Gjertsen, A., Ljunggren, E. and Pettersen, L.T. (1999) *Blir det arbeidsplasser av dette da, jenter?*, Nordland Research Institute Report 13, Bodø.

Birkelund, G.L. and Petersen, T. (2003) Det norske likestillingsparadokset – kjønn og arbeid i et velferdssamfunn, in I. Frønes and L. Kjølsrød (eds.) *Det norske samfunn*, fourth edition, Gyldendal Akademisk, Oslo.

Blekesaune, Arild, Haney, Wava G. and Haugen, Marit S. (1993) On the question of the feminization of production on part-time farms: Evidence from Norway, *Rural Sociology* 58, pp.111-129.

Countryside Agency (2002) *Rural Proofing 2001/2*, Countryside Agency, Cheltenham.

Eidheim, Frøydis (1993) *Sett Nordfra. Kulturelle aspekter ved forholdet mellom sentrum og periferi*, Universitetsforlaget, Oslo.

Ellingsæter, Anna Lise (1995) Kjønn, deltidsarbeid og fleksibilitet iI arbeidsmarkedet: Det norske eksemplet, in Dag Olberg (ed.) *Endringer i arbeidslivets organisering*, Fafo, Oslo.

Ellingsæter, Anna Lise and Solheim, Jorun (2002, eds.) *Den usynlige hånd? Kjønnsmakt og moderne arbeidsliv*, Gyldendal Norsk Forlag a/s Akademisk, Oslo.

Foucault, Michel (1999) *Diskursens orden*, Spartacus, Oslo.

Giddens, Anthony (1990) *The Consequences of Modernity*, Polity, Cambridge.

Hansen, J.K. (1998) *Why do young people leave fishing communities in coastal Finnmark, North Norway?*, paper presented at the International Geographical Union Commission on Population and the Environment session on 'Population and Environment in the Atlantic Islands', 30 August 1998, Lisbon.

Haraway, Donna (1995) Situerte kunnskaper, in Kristin Asdal, Brita Brenna, Ingunn Moser and Nina Refseth (eds.) *En kyborg til forandring – nye politikker i moderne vitenskaper og teknologier*, TMV-senteret, Oslo.

Harding, Sandra (1991) *Whose Science? Whose Knowledge? Thinking from Women's Lives*, Open University Press, Buckingham.

Kommunal- og Regional Departmentet (2001) *Stortingsmelding nr 34 (2000-2001), Om distrikts- og regionalpolitikken*, Oslo [Ministry of Local Government and Regional Development. White Paper on Local and Regional Development in Norway].

Kylhammar, Martin (2003) *Kommunikation, kunnskap, makt*, in Martin Kylhammar, and Jean-Francois Battail (eds.) *På Väg Mot En Kommunikativ Demokrati? Sexton*

Humanister Om Makten, Medier, Carlsson Bolförlag, Sverige.

Likestillingssenteret (2003) *Likestillingsbarometeret 2003*, Likestillingssenteret, Oslo.

Limstrand, I. (1996) *Mellom lokalsamfunn og storsamfunn*, Hovedoppgave, Norsk Teknisk Naturvitenskaplig Universitetet Geografisk Institutt, Trondheim.

Lotherington, A.T. (2002) *Ikke for kvinnenes skyld... En analyse av kvinnerettet distriktspolitikk I Norge 1980-2000*, Norut Samfunn, Tromsø.

Lundgren, Eva (1985) *I Herrens vold*, J.W. Cappelen A/S, Oslo.

Makt- og demokratiutredningen, Nou 19, 2003.

Massey, D.B. (1991) A global sense of place, *Marxism Today*, June, pp.24 –29.

Nordland Fylkeskommune (2003) *Kvinnesatsing i Nordland – Utredning om kvinner of likestilling*, Nordland County, Bodø.

Otterstad, O. and Jentoft, S. (1994, eds.) *Leve kysten? Strandhogg i fiskeri-Norge*, Ad Notam Gyldendal, Oslo.

Pettersen, L.T. (1994) Hovedsaken er at kjerringa er i arbeid: husholdsstrategier i fiskerikrisen, in O. Otterstad and S. Jentoft (eds.) *Leve kysten? Strandhogg i fiskeri-Norge*, Ad Notam Gyldendal, Oslo, pp.65-76.

Pettersen, L.T. (1999) *Likestilling, kvinner og kjønn i samfunnsplanleggingen*, Nordisk Arkitekturforskning 2/99, Oslo.

Rossvær, Viggo (1994) Tradisjon og modernitet: Kan utkantene være moderne?, in O. Lian (ed.) *Mellom tradisjon og modernitet*, Universitetet i Tromsø, Institutt for samfunnsvitenskap Stensilserie A nr. 76, Tromsø.

Scholten, C. (2004) Partnerships for regional development and the question of gender equality, in H. Buller and K. Hoggart (eds.) *Women in the European Countryside*, Ashgate, Aldershot, pp.103-122.

Sivertsen, B.B., van Schaik, N., Friis, B. and Stemland, M. (2003) *Hvordan kan veiledning legge til rette for tverrfaglig samarbeid i en organisasjon i endring i en utkantkommune?*, Høgskolen i Hedmark, Elverum.

Skjønsberg, E. (1995) *Omsorgsrasjonalitet. Fremtidens fornuft. Det alternative bibliotek*, Universitetsforlaget, Oslo.

Sørlie, K. (2000) *Klassiske analyser: flytting og utdanning belyst i et livsløps- og kohortperspektiv*, Norwegian Institute for Urban and Regional Research NIBR-notat 121, Oslo.

Thordarson, J. (2003) *Fiskerinæringen, en kunnskapsbasert næring?* Rapport fra Nordisk Ministerrådskonferanse, Fun – fisk og ungdom i Norden – om bedre rekruttering og utdanning i fiskerinæringen, 25–26 okt 1999, København.

Vikan, S.T. (2000) *Kvinner og menn i Norge 2000*, Statistisk Sentralbyrå, Oslo-Kongsvinger.

Widerberg, K. (1992) Teoretisk verktøykasse – angrepsmåter og metoder, in A. Taksdal and K. Widerberg (ed.) *Forståelser av kjønn i samfunnsvitenskapenes fag og kvinneforskning*, Ad Notam Gyldendal, Oslo, pp.285-299.

Woie Berg, B., Fylling, I. and Stemland, M. (1993) *Kvinneuniversitetet Nord*, Northern Feminist University, Nordfold.

Index

Printed in the United States
by Baker & Taylor Publisher Services